An Introduction to Panel Data QCA in R

In the last few years, Qualitative Comparative Analysis (QCA) has become one of the most important research approaches in social science. This has encouraged researchers to apply QCA, to analyze cross-sectional and panel data, leading to the development of a variety of cross-sectional and panel data QCA models.

This book compares four different panel data QCA models: *Cluster QCA, Multiple Sub-QCA, Remote-Proximate Panel,* and *Relevant Variation Panel.* It starts by introducing QCA as a research approach, then discusses the assumptions, and steps in a QCA research process. It then applies these assumptions and steps to demonstrate each of the 4 afore-mentioned panel data QCA models. Each chapter also provides a step-by-step guide, that researchers can follow while building any of these 4 panel data QCA models. Finally, it compares the strengths and weaknesses of each of these models and suggests scenarios where researchers can apply them. This book is supplemented by materials like datasets and codes, available at the end of each chapter, and online on Harvard Dataverse. This book can be used as a textbook for introductory and advanced courses on panel data QCA.

Preya Bhattacharya was a postdoctoral fellow at the Sam Nunn School of International Affairs, Georgia Institute of Technology. She has published articles in the *International Journal of Social Research Methods* and *Methodological Innovations.* Along with panel data QCA, her research and teaching interests include microfinance, economic participation of women, and post-conflict economic development.

An Introduction to Panel Data QCA in R

Preya Bhattacharya

CRC Press
Taylor & Francis Group
Boca Raton London New York

CRC Press is an imprint of the
Taylor & Francis Group, an **informa** business

Designed cover image: © Preya Bhattacharya

First edition published 2024
by CRC Press
2385 NW Executive Center Drive, Suite 320, Boca Raton FL 33431

and by CRC Press
4 Park Square, Milton Park, Abingdon, Oxon, OX14 4RN

CRC Press is an imprint of Taylor & Francis Group, LLC

ISBN: 9781032470962 (hbk)
ISBN: 9781032471099 (pbk)
ISBN: 9781003384595 (ebk)

DOI: 10.1201/9781003384595

Typeset in Times
by Apex CoVantage, LLC

Contents

Contents

Introduction

1

Qualitative Comparative Analysis (QCA) was originally developed by Charles G. Ragin, in 1987, as a comparative case study approach, in his book *The Comparative Method* (1987, 2014). Since then, QCA has been applied in various disciplines of social science, including political science, international relations, sociology, and management studies, among others (Mello 2021, 4–10; Oana, Schneider, and Thomann 2021, 18–21), to analyze cross-sectional as well as panel data.

In this book, I discuss four different approaches toward analyzing panel data in QCA (Oana, Schneider, and Thomann 2021, 164; Verweij and Vis 2021, 95–111). These are:

1. Cluster Analysis in Panel Data QCA (Chapter 5: *Cluster QCA*)
2. Separate cross-sectional QCA models (Chapter 6: *Multiple Sub-QCA*)
3. Two-Step Panel data QCA (Chapter 7: *Remote-Proximate Panel*)
4. Set-Theoretic Approaches to Change (Chapter 8: *Relevant Variation Panel*)

OUTLINE OF THE BOOK

In chapter 2, I outline the assumptions and steps a QCA research process. In chapter 3, I discuss the importance of panel data in QCA, and the different types of panel data QCA models. In Chapter 4, I briefly outline and demonstrate one of the most important steps in QCA: Calibration.

Between chapters 5 to 8, I demonstrate each of the 4 panel data QCA models. To demonstrate these approaches, I have applied a portion of my own research data, on microfinance and national economic participation of women (Bhattacharya, 2020).

In my research, I have analyzed how domestic political policies influence the success of internationally-financed microfinance programs. I have

DOI: 10.1201/9781003384595-1

compared three former Yugoslavian countries, Bosnia-Herzegovina, Croatia, and Montenegro, between the years, 1999–2007, and theorized that microfinance will be more effective, in countries with strong government regulations. I have built my causal analysis model, based on Sen's Theory of Development as Freedom (1999), and my own case-study data. I have tabulated my necessary and sufficient conditions, and my outcome variables, in chapter 4, Table 4.7.

In Chapter 5, I demonstrate Garcia-Castro and Ariño's (2016) panel data QCA model. I have termed this approach as *Cluster QCA*.

In Chapter 6, I apply Verweij and Vis's (2021) approach to analyzing panel data in QCA, based on establishing separate cross-sectional QCA models. I have termed this approach as *Multiple Sub-QCA*.

In Chapter 7, I demonstrate Schneider and Wagemann (2012) and Schneider's (2019) two-step QCA approach. I have termed this approach as *Remote-Proximate Panel*.

In Chapter 8, I demonstrate Ragin and Fiss (2019) Set-Theoretic Approaches to Change approach (Ragin and Fiss 2019). I have termed this approach as *Relevant Variation Panel*.

Though the basic assumptions of each of these models are similar to one another, there are a few additional assumptions, that I have discussed in each chapter separately (chapters 5–8). I have included appendix materials at the end of each chapter, and these are also available on Harvard Dataverse, https://doi.org/10.7910/DVN/F0PT9C

I conclude this book, by comparing the four approaches and offering suggestions on when to implement these four different types of panel data QCA models.

Assumptions and Steps of a QCA research process

2

Chapter outline:

- What is Qualitative Comparative Analysis (QCA)?
- Importance
- Assumptions
- Analyzing causality in QCA
- The steps in a QCA research process
- The six types of panel data QCA models
- The aim of this book

In this chapter, I discuss the origin, development, assumptions, and how to analyze causality in QCA.

WHAT IS QUALITATIVE COMPARATIVE ANALYSIS?

As mentioned in Chapter 1, Qualitative Comparative Analysis (QCA) was originally developed by Charles C. Ragin, in his book *The Comparative Method* (1987), as a comparative case-oriented research method, based on the principles of Boolean algebra (Boole 1847, 1854) and set theory (Zadeh 1965).

DOI: 10.1201/9781003384595-2

Ragin applied these principles to test causality with the help of necessary and sufficient conditions, amongst comparable cases in the field of social sciences, with (Mill 1873; Lijphart 1971; Ragin 1987). In the last few years, QCA has been applied to various social science subjects such as political science, sociology, economics, and gender studies, among others.

IMPORTANCE OF QCA

As a comparative research approach, QCA is important for three main reasons (Goertz 2017; Oana, Schneider, and Thomann 2021, 180–200):

First, researchers can apply QCA to analyze a *small to medium number of cases*, from the *same level or various levels of analysis*.

Second, QCA can help us analyze *how a factor or combination of factors impact an outcome*, by identifying causal pathways, through the testing of necessary and sufficient conditions.

Third, researchers can apply QCA, to *test and build new theories*, and *understand contextual variation*.

THE MAIN ASSUMPTIONS OF QCA

There are six main assumptions of QCA. These are:

First, in QCA, independent variables are known as conditions, and dependent variables are known as outcomes (Ragin 1987, XXIII).

Second, in QCA, we can study the effect of multiple conditions by creating new conditions through conjunction (logical AND) or disjunction (logical OR) (Schneider and Wagemann 2012, 42–47).

Third, QCA emphasizes equifinality or multiple causation (Schneider and Wagemann 2012, 78–79). According to this assumption, there might be many different pathways to the same outcome. Each of these pathways is joined through addition, known as logical OR. Each pathway might also consist of INUS or SUIN conditions (Schneider and Wagemann 2012, 79–80). INUS conditions are "Insufficient but Necessary part of a Condition which is itself Unnecessary but Sufficient for an Outcome," usually found while testing for sufficient conditions. SUIN conditions are "Sufficient but Unnecessary part of a Factor that is Insufficient but Necessary for an Outcome," found while testing for necessary conditions.

Fourth, in QCA, researchers analyze causality by testing for necessary and sufficient conditions (Schneider and Wagemann, 2012; Dusa, 2022). A *necessary condition* is a condition that is always present when the outcome occurs; the outcome cannot happen in the absence of the condition, and the condition is a superset of the outcome. A *sufficient condition* is a condition that can cause the outcome, but is not the only factor that can do so; the presence of the condition does not lead to the absence of the outcome, and it is a subset of the outcome. I discuss the process of testing for necessary and sufficient conditions in QCA in further detail in what follows.

Fifth, necessary conditions can be single conditions, but they can also be higher-order concepts, known as SUIN conditions (Schneider and Wagemann 2012, 80). We can create these higher-order concepts theoretically, or by applying the superSubset() function (Dusa 2022, 136–137).

Similarly, sufficient conditions in QCA can be single conditions. But we can also have sufficient conditions combining to form a pathway. This pathway is known as conjunction, and each sufficient condition in a pathway is known as conjunct, joined by AND (*) (Oana, Schneider, and Thomann 2021, 184). Each conjunct is also known as an INUS condition (Insufficient but Necessary part of a condition which is itself Unnecessary but Sufficient for an Outcome) (Schneider and Wagemann 2012, 79). Since QCA is based on equifinality or multiple causation, there can be multiple pathways that are sufficient for an outcome. Each of these pathways is joined by OR (+).

Fifth, QCA is based on the set-theoretic assumptions of asymmetric relations (Schneider and Wagemann 2012, 81–83), meaning that the reasons for the occurrence of an outcome are different from the reasons for the non-occurrence of an outcome (Schneider and Wagemann 2012, 81–83).

As a result, while testing for necessary conditions, we need to test the presence of each necessary condition on the presence of an outcome separately from the absence of each necessary condition on the presence of an outcome (Schneider and Wagemann 2012, 71). While testing for sufficient conditions, we need to test the reasons for the occurrence of an outcome, separately, from the reasons for the non-occurrence of an outcome (Ragin 1987; Schneider and Wagemann 2012, 59).

Sixth, in QCA, the unit of analysis can be a data category (observational units) or a theoretical category (explanatory units) (Ragin 1987, 7–9). Thus, researchers can apply given populations (cases) at the micro-, meso-, or macro-levels, to analyze their research questions, or researchers can select/create cases based on their research questions and the concepts they want to analyze (Ragin and Becker 1992, 2009, 1–17). These cases should have certain common or shared characteristics, which will help researchers to compare and analyze.

Along with these assumptions, another important term that is frequently applied in QCA is *calibration*. In QCA, since each variable is regarded as a set, each case has a degree of membership in that set (Schneider and Wagemann 2012, 32). To determine this degree of membership, each variable (conditions and outcome) has to be calibrated separately (Schneider and Wagemann 2012, 32). I discuss calibration in detail in the next section.

I have listed the most important notations/symbols and terms used in QCA in Tables 2.1 and 2.2.

TABLE 2.1   Symbols of QCA

QCA OPERATORS	MEANING	DENOTED BY
AND	Takes the minimum of two values	Set theory: Intersection (X ∩ Y) QCA: Multiplication
OR	Takes the maximum of two values	Set theory: Union (X ∪ Y) QCA: Addition (+)
Negation	Absence of a condition/ outcome	Set Theory: 1-Membership of Set QCA: Complement (~)
Subset	Sufficient Condition	Set theory: Subset (⊂) QCA: x -> y
Superset	Necessary Condition	Set theory: Superset (⊃) QCA: x <- y

Source: Adapted from the Southern California QCA Workshop (Ragin and Fiss 2019), & Schneider and Wagemann (2012, 54)

TABLE 2.2   QCA Terminologies

CONVENTIONAL TEMPLATE	QCA TEMPLATE
Variables	Sets
Measurement	Calibration
Independent Variables	Conditions
Dependent Variables	Outcomes
Given populations	Constructed populations
Correlations	Set-theoretic relations (superset/subset)
Correlation matrices	Truth tables
Net effects	Causal Recipes (chemical causation)

Source: Adapted from Ragin (1987, XXIII) and the Southern California QCA Workshop (Ragin and Fiss 2019)

ANALYZING CAUSALITY IN QCA

In QCA, researchers analyze causality with the help of necessary and sufficient conditions.

Conceptually, a condition is defined as necessary if it must be present for the outcome to occur (Ragin 1987, 99). A condition (X) is necessary for Y, if X is always present when Y occurs, Y does not occur in the absence of X, and Y is a subset of X (Dusa 2022, 113–114). X as a necessary condition means that X is the only factor that can cause Y (Dusa 2022, 113–114). The statement "X is necessary for Y" can be written as "X <= Y" (Dusa 2022, 113–114). In QCA, a condition X is necessary for Y, if its consistency value is at least equal to 0.9 (Schneider and Wagemann 2012, 143). After measuring the consistency of a condition, a researcher can also measure the coverage value, which is a measure of how trivial or relevant a necessary condition X is for an outcome Y (Dusa 2022, 124–133).

In QCA, if X is a necessary condition for Y, then the absence of condition X cannot lead to the presence of outcome Y (Schneider and Wagemann 2012, 71; Mello 2021, 57). To understand necessity, we need to test the presence of condition X, separately from the absence of condition X, as seen in Table 2.3 (Schneider and Wagemann 2012, 71; Mello 2021, 57).

So, if X is necessary for Y, then there should be cases where X and Y are both present (Option A, Table 2.3). But, there shouldn't be cases where X is absent, but Y is present (Option C, Table 2.3). Graphically, if a condition X is necessary for Y, then all or most of the data points, should fall below the diagonal.

In QCA, researchers can also analyze causality with the help of sufficient conditions. A sufficient condition is a condition that can produce the outcome, but is not the only condition to do so (Ragin 1987, 99). As an example, X is a sufficient condition for Y, if X is present when Y occurs, X does not occur in the absence of Y, and X is a subset of Y (Dusa 2022, 143). In the testing of sufficiency, X can be a single condition or a combination of conditions (*) and is denoted by "X => Y". For a sufficient condition, the consistency score should be at least 0.75, along with high coverage (Schneider and Wagemann 2012, 129). So, if X is sufficient for Y, then there should be cases where X and Y are both present (Option A, Table 2.4). But, there shouldn't be cases where X is present, but Y is absent (Option B, Table 2.4). Graphically, if a condition X is suffficient for Y, then all or most of the data points, should be above the diagonal.

To test for sufficiency, a researcher needs to apply the steps discussed below. First, form a truth table that displays all logically possible combinations of conditions and their impacts on an outcome. In a truth table, the columns represent the different variables/sets, and the rows represent the number of logically possible combinations of values for the causal variable/

TABLE 2.3 Necessary Condition and Outcome

OPTIONS	CONDITION (X)	OUTCOME (Y)	X AS A NECESSARY CONDITION FOR Y
A	1	1	Allowed, many cases
B	1	0	Allowed, not relevant, some cases
C	0	1	Not allowed, no cases
D	0	0	Allowed, not relevant, not important

TABLE 2.4 Sufficient Conditions and Outcome

OPTIONS	CONDITION (X)	OUTCOME (Y)	X AS A SUFFICIENT CONDITION FOR Y
A	1	1	Allowed, many cases
B	1	0	Not allowed, no cases
C	0	1	Allowed, not relevant, some cases
D	0	0	Allowed, not relevant/not important

set (Ragin 1987, 87–89). As an example, if k is the total number of conditions and 2 represents the presence or absence of condition, then the total number of rows in a truth table is equal to 2^k (Ragin 1987, 87–89; Schneider and Wagemann 2012, 92–104). As shown in Table 2.5, the last column represents the outcome column (Y). Depending on the values of the input, each row is assigned an output value in column Y (Ragin 1987, 88).

In QCA, there are five main types of outcome results: absence of outcome (0), presence of outcome (1), indeterminate or don't care (-), Logical Remainders (L/R), and Contradictory Outcome (C) (Rihoux and Ragin 2009, 44).

Indeterminate outcomes (-) are rows where the outcome could not be decided (Rihoux and De Meur 2009, 44–48). As the aim of a QCA model is to explain the outcome, a researcher should resolve rows with indeterminate outcomes, either by adding including new case studies and/or variables, or by removing case studies and/or variables (Rihoux and De Meur 2009, 44–48).

Logical Remainders, denoted by L/R, are the logically possible combinations of conditions that could not be observed among the cases studied (Rihoux and De Meur 2009, 44–48). A researcher has three main options to resolve these logical remainders (Rihoux and De Meur 2009, 44–48). A researcher can remove logical remainders (conservative/complex solution), include but not explain logical remainders (parsimonious solution), or include and explain logical remainders (intermediate solution) (Rihoux and De Meur 2009, 44–48).

A **Contradictory Outcome**, denoted by C, occurs when the same combination of conditions leads to an outcome's absence in one case and presence in another (Rihoux and De Meur 2009, 44–48). A researcher needs to resolve these contradictory outcomes before logically minimizing the truth table (Rihoux and De Meur 2009, 44–48). To resolve contradictory outcomes, a researcher might have to add/remove conditions or cases studied (Rihoux and De Meur 2009, 44–48). A researcher can also apply the pof() function, to analyze whether X is sufficient for the presence or absence of Y, by interpreting the PRI score, which should be at least greater than 0.5, and maximum of 1.0 (Dusa 2022, 150–151).

The second step is to simplify the results of a truth table through Boolean minimization which helps us identify redundant conjuncts and logically redundant prime implicants (Ragin, 1987: 93–98; Oana, Schneider, and Thomann 2021, pp. 112–114). Boolean minimization helps us remove redundancies from the truth table.

The third step is to identify and address logical remainders (with outcome, ?) (Oana, Schneider and Thomann, 2021, pp. 122–130). This step is knows as standard analysis, and it creates three types of solutions: conservative/complex, most parsimonious, and intermediate (Oana, Schneider and Thomann, 2021, pp. 122–130).

The fourth step is to create enhanced standard analysis solutions by identifying and removing untenable assumptions, logical remainders which are contradictory simplifying, contradicts claims of necessity, and are implausible remainders (Oana, Schneider and Thomann, 2021, pp. 131–140). There are three different types of enhanced standard analysis solutions: enhanced conservative, enhanced parsimonious, and enhanced intermediate. In QCA, researchers are always encouraged to focus on the enhanced intermediate solution for further analysis (Dusa, 2019: 139–193; Schneider and Wagemann, 2012). This enhanced standard analysis solution is the simplest combination of conditions that can cause an outcome (Ragin 1987, 93–98; Dusa 2022, 179–180).

Hence, the main steps involved in testing for sufficiency are: first, form a truth table; second, logically minimize the truth table to remove redundant conjuncts and logically redundant prime implicants; third, create the standard analysis solutions through the conservative, parsimonious, and intermediate solution; fourth, identify untenable assumptions and remove them to create the enhanced standard analysis solutions, enhanced conservative, enhanced parsimonious, and enhanced intermediate (Dusa 2022, 141–214; Oana, Schneider, and Thomann 2021, 86–140). In the next section, I have discussed the main steps in a QCA research process, before and after testing for necessary and sufficient conditions.

TABLE 2.5 Hypothetical Truth Table

	CONDITIONS		OUTCOME
ROWS	A	B	Y
1	0	0	0
2	0	1	0
3	1	0	1
4	1	1	1

Source: Adapted from Ragin (1987, 88)

In Table 2.5, each of the logically possible combinations (0,0), (0,1), (1,0), (1,1) is assigned a separate row. Depending on the value of each possible combination, each row is assigned an outcome value (Y column).

THE STEPS IN A QCA RESEARCH PROCESS

In QCA, researchers are usually encouraged to follow five main steps (Rihoux and Lobe 2009, 230–237): gain case study knowledge, specify a theory, calibrate each set, test for necessary conditions, and test for sufficient conditions. If we have adopted a theory-testing approach, we also need to test the applicability of the theory through theory evaluation. In this section, I discuss each of these five steps in greater detail.

Step 1: Gain case study knowledge

In a QCA research process, the first step is to gain an in-depth knowledge of cases (Rihoux and Lobe 2009, 230–231). This case study knowledge helps a researcher identify conditions that can be included in the data analysis model, and interpret the data analysis results in the context of the cases studied (Rihoux and Lobe 2009, 230–231). A researcher can gain this case study knowledge through various case study methods like historical narrative, content analysis, and/or process-tracing.

Step 2: Specify a theory

Along with case study knowledge, theories play a key role in deciding the conditions that should be selected for building a QCA model. Hence, the second step is to review the existing literature to establish whether any existing theory/ies can be applied to analyze the research question or topic studied. Researchers can apply these prior existing theory/ies to decide which conditions to include in the data analysis model. Thus, the final data analysis model is built by including variables from the case study knowledge, as well as from existing theories.

Scholars including Rihoux and Lobe (2009, 232) have argued that there needs to be a balance between the number of conditions and cases studied, as the number of possible combinations increases exponentially with the number of conditions included in the data analysis model. So, the number of conditions in a QCA research model should be 4–6/7 for every 10–40 cases (Rihoux and Ragin 2009, 28). To maintain this balance, researchers need to prioritize variables from the case study or the theory, depending on whether it is a theory-building or a theory-testing research process (Rihoux and Lobe 2009, 232).

If it is theory-building, then primary importance is given to case study knowledge (Rihoux and Lobe 2009, 232). If it is theory-testing, then primary importance is given to the variables from the selected theory (Rihoux and Lobe 2009, 232).

We can also create macro-conditions, to reduce the number of conditions, by combining variables through fuzzyor(), or fuzzyand() (Schneider and Wagemann 2012, 292–293; Ragin and Fiss, Southern California QCA Workshop 2019). Or, apply the two-step process by testing for remote conditions as SUIN necessary conditions, and then testing for remote-proximate conditions (Schneider and Wagemann 2006; Schneider 2019).

Step 3: Calibration

Once the conditions have been selected, the third step in a QCA research process is to calibrate each variable (conditions and outcome). Variables in a QCA model are defined as sets (Ragin 1987, XXIII). Each case/data value/observation is assigned a membership to the set (Ragin 1987, XXIV, 2008, 71–72). This process of assigning set-membership is known as calibration Ragin 1987, XXIV).

According to Ragin, calibration is important because it helps to make measurements interpretable in the context of the cases studied, that is, closer to social science reality (Ragin 1987, XXIV). This calibration criterion is usually based on established social knowledge external to the data or the researcher's own internal case study knowledge (Schneider and Wagemann 2012, 32–41; Dusa 2022, 75–109; Goertz and Mahoney 2012, 157; Oana, Schneider, and Thomann 2021, 27–47).

In the calibration stage, the number of thresholds established depends on the type of QCA chosen. There are two main types of QCA: crisp-set QCA and fuzzy-set QCA. For crisp-set QCA, there are two main qualitative states, 0 (non-membership) and 1 (membership). The number of thresholds in crisp-set QCA is 2. Fuzzy-set QCA usually ranges from 0 to 1. Since thresholds tend to represent qualitative states, there are three main qualitative states in fsQCA: 0 (non-membership), 0.5 (partial membership), and 1 (membership) (Schneider and Wagemann 2012, 28). Depending on the type of data, there are three main ways of establishing these thresholds in fsQCA.

If it is qualitative data, the researcher applies their own case study knowledge through content analysis or process-tracing and establishes thresholds as suggested by Basurto and Speer (2012) and Oana, Schneider, and Thomann (2021, 46).

If it is quantitative data, the researcher can apply case study knowledge to establish the qualitative thresholds through a theory-guided calibration process (Oana, Schneider, and Thomann 2021, 38–39). Researchers can also apply available social science external criteria to establish the qualitative thresholds of 0, 0.5, and 1. This is known as Ragin's Direct Calibration Method for fuzzy-sets (Ragin 2008; Oana, Schneider, and Thomann 2021, 42–43). These qualitative thresholds need to be established for each of the variables separately. Once the qualitative thresholds have been established, researchers need to specify quantitative gradings or difference-in-degree in between the qualitative thresholds (Dusa 2022, 88–106). To specify these differences in degree, we can create three- to seven-value fuzzy-sets or continuous fuzzy-sets, tabulated in Tables 2.6 and 2.7. We do not need to assign cases to each of these membership scores; in fact, we can create fuzzy-set membership scores with unequal intervals based on theory and case study knowledge (Schneider and Wagemann 2012, 28–29).

If the data is quantitative and literature is not available to establish the qualitative thresholds, the researcher can apply Ragin's Indirect Method of Calibration (Dusa 2022, 106–108). According to Ragin's Indirect Calibration Method, a researcher first needs to determine the distribution of data, and depending on the distribution they can calibrate the data with the help of regression models like fractional polynomial, beta, or binomial (Oana and Schneider 2018; Dusa 2022, 106–108).

Ragin's Direct Calibration and Indirect Calibration Methods are mostly used for interval data, but if we have categorical data with more than two categories, we can calibrate it into a crisp-set or a totally fuzzy and relative set (TFR) (Dusa 2022, 109).

Along with these aforementioned calibration techniques, we also need to keep in mind the "principle of unimportant variation" while calibrating a set, and calibrate a set based on relevant qualitative/quantitative variation (Goertz and Mahoney 2012, 144–145).

TABLE 2.6 The Different Types of Fuzzy-Sets (Table based on Ragin 2000, 156; Ragin 2008, 31)

THREE-VALUE FUZZY-SET	FOUR-VALUE FUZZY-SET	FIVE-VALUE FUZZY-SET	SIX-VALUE FUZZY-SET	SEVEN-VALUE FUZZY-SET	CONTINUOUS FUZZY-SET
1=fully in	1=fully in	1=fully in	1=fully in	1=fully in	1=fully in
			0.8=mostly but not fully in	0.83=mostly but not fully in	
	0.67=more in than out	0.75=more in than out	0.6=more or less in	0.67=more in than out	0.5< × <1: More in than out
0.5= neither fully in nor out		0.5=crossover point, neither in nor out		0.5=crossover point	0.5=crossover point neither in nor out
	0.33=more out than in	0.25=more out than in	0.4=more or less out	0.33=more or less out	0 < × <0.5: More out than in
			0.2=mostly but not fully out	0.17=mostly but not fully out	
0=fully out	0=fully out	0=fully out	0=fully out	0=fully out	0=fully out

TABLE 2.7   Fuzzy-Set Membership Scores (Based on Schneider and Wagemann 2012, 29)

FUZZY-VALUE	QUALITATIVE THRESHOLDS
1	Fully in
0.9	Almost fully in
0.8	Mostly in
0.6	More in than out
0.5	Crossover, neither in nor out
0.4	More out than in
0.2	Mostly out
0.1	Almost fully out
0	Fully out

Step 4: Test for necessary and sufficient conditions

After all the variables have been calibrated into sets, we can then start testing for necessary and sufficient conditions. In a panel data QCA, the testing for necessary and sufficient conditions can help us understand how the impact of a condition has changed over time and across cases/countries studied.

Step 5: Theory evaluation in QCA

If we are applying a theory-testing research approach, then the last step in a QCA research process is to test the applicability of the theory, in the context of the research question and the cases studied. In QCA, this theory-testing is known as theory evaluation, and the aim is to identify the scope or boundaries of the theory tested (Ragin 1987, 2014, 11–12, 84, 118–121; Schneider and Wagemann 2012, 297–305).

In a theory-testing QCA model, competing theories can be joined with a logical OR (+) to find a simplifying solution (Dusa 2022, 216). At the end of the Boolean analysis, the researcher can contrast the results with those of the theoretically based expectations (Ragin 1987, 2014, 118–121). For example, the results of the Boolean analysis might show that there are a few cases that match the theoretical expectation, while there are cases that do not match the theoretical expectation at all.

As explained by Ragin (1987), if T is the theory and R is the result, then the options in theory evaluation are (2014, 118–121):

- TR -> subset of causal conditions that were both hypothesized and found
- tR -> subset of causal conditions that were found but not hypothesized
- Tr -> subset of causal combinations hypothesized but not found
- tr -> subset of causal combinations not hypothesized and not found

These options or clusters of cases can then be applied by the researcher to extend and build a new theory (Schneider and Wagemann 2012, 304). Each of the steps discussed above, can be applied for cross-sectional and panel data QCA models. But, in this book, I have demonstrated how to apply these steps only in the case of panel data QCA models. I will start by discussing the six main types of panel data QCA models below.

THE SIX TYPES OF PANEL DATA QCA MODELS

Through panel data QCA, researchers aim to analyze **how social structures/ macrosocial units have changed over time and across cases** (Ragin 1987, 2014, 6–7, 11). There are six main ways to do this:

First, we can follow Caren and Panofsky's *temporal QCA* (TQCA) model, to understand the order in which conditions occur, that is, whether condition A occurs before condition B and whether this sequence influences the impact of conditions 2005.

Second, we can apply Baumgartner's *Coincidence Analysis* model (2009) to understand the common structure among the conditions and the outcome by identifying causal chains (Oana, Schneider, and Thomann 2021, 166–170).

Third, we can adopt Schneider and Wagemann's (2006) and Schneider's (2019) *two-step* model to analyze remote and proximate causes (759–762). Remote causes are factors that are mostly structural in nature, have not changed over time, and occurred a long time before the outcome happened (Schneider and Wagemann 2006, 759–762), while proximate causes are contextual factors, have changed over time, and occurred closer in time to the outcome studied (Schneider and Wagemann 2006, 759–762). According to the assumptions of this two-step model, a researcher should first test for the remote causes as necessary SUIN conditions (first step), and then include the relevant SUIN conditions along with the remaining sufficient conditions in the final model, to test for the relevance of proximate causes as sufficient conditions (second step) (Schneider 2019).

In case we have not been able to identify the sequencing or order in which conditions occur, or differentiate between the remote and proximate factors common structure, we can apply *three additional* models (as discussed below), to understand how the impacts of conditions have changed over time and across cases studied.

We can *establish separate QCA* models at different time intervals, and then analyze how time influences a change in the impact of these variables (Verweij and Vis 2021, 100–105).

We can follow Garcia-Castro and Ariño (2016) to set up a *panel data QCA* model, with a column identifying the clustering element (for example, time in years) and a second column identifying the level of data collection.

We can also apply *Ragin's Set-Theoretic Approaches to Change* by analyzing the qualitative change in variables over the time period studied (Ragin and Fiss, Southern California QCA Workshop 2019).

THE AIM OF THIS BOOK

In this book, I will demonstrate the third (Schneider and Wagemann 2006; Schneider 2019), fourth (Verweij and Vis 2021, 100–105) fifth (Garcia-Castro and Ariño 2016), and sixth models (Ragin and Fiss 2019), of analyzing panel data QCA.

I have termed these models as *Cluster QCA* (Chapter 5, Garcia-Castro and Ariño 2016); *Multiple Sub-QCA* (Chapter 6, Verweij and Vis 2021)*; Remote -Proximate Panel* (Chapter 7; Schneider 2019)*; Relevant Variation Panel* (Chapter 8; Ragin and Fiss 2019). In the next few chapters, I will apply my research data to demonstrate and build each of these models, and then compare the data analysis results from each, to address two main questions:

- First, is there any difference in data analysis results?
- Second, if there is a difference in data analysis results, then which approach should be adopted and under what circumstances?

Panel data and QCA

3

Chapter outline:

- What is panel data QCA?
- Importance of panel data QCA
- QCA packages and panel data QCA in R

WHAT IS PANEL DATA QCA?

Initially, scholars usually applied QCA to study cross-sectional data. However, in the last few years, scholars have started to apply QCA to study panel data as well, developing a variety of models to study how configurations change over time. As stated by Ragin (1987, 2014, 6–7, 11), researchers can apply QCA to analyze how macrosocial units have changed over time (p. 6).

To do this, scholars have developed a variety of QCA models. The assumptions of most of these models are similar to cross-sectional QCA, with a few differences, which I have discussed in the individual chapters.

Along with these assumptions, researchers need to structure their panel data QCA differently, as compared to cross-sectional data in QCA. As an example, for *Cluster QCA* (Chapter 5) and *Remote-Necessary & Proximate-Sufficiency* (Chapter 7), researchers need to structure their data in the long format, while the data for *Relevant Variation Panel* (Chapter 8), needs to be arranged in the wide format.

DOI: 10.1201/9781003384595-3

IMPORTANCE OF PANEL DATA QCA

Panel data QCA is important because:

- It can help us understand whether the impact of condition changes over time and across cases.
- Analyze how contextual factors have evolved over time.
- Like cross-sectional QCA, researchers applying panel data QCA can study a smaller number of cases.
- Researchers can also include data from various levels of analysis, such as micro, meso, and macro, to build a panel data QCA model.

PANEL DATA QCA IN R

To analyze panel data QCA models in RStudio, researchers need to install and run two main packages. These are the SetMethods package (Oana and Schneider 2018) and the QCA package (Dusa 2019). Researchers can apply the install.packages() and library() functions, to download and run these packages.

Data calibration

4

Chapter outline:

- Introduction
- Calibrating crisp sets
- Calibrating fuzzy-sets
- Applying theory-guided calibration to create fuzzy-sets
- Applying Ragin's Direct Calibration approach
- Ragin's Indirect Calibration approach
- Conclusion
- Appendix

INTRODUCTION

In this chapter, I will demonstrate how to calibrate quantitative data into crisp and fuzzy-sets. Alhough I have not discussed the process of calibrating qualitative data or creating multi-value sets in this book, researchers are welcome to refer to the works of Basurto and Speer (2012), Schneider and Wagemann (2012), Dusa (2022), Oana, Schneider, and Thomann (2021), and Mello (2021).

In the next section, I will first describe how to create a crisp-set, and then how to create fuzzy-sets, by applying Theory-Guided Calibration, Ragin's Direct Calibration, and Ragin's Indirect Calibration methods. I will apply the dataset from Chapter 5 to demonstrate these calibration techniques.

Since the calibration results depend on the data distribution, I have calibrated all the variables separately for every panel data QCA model discussed in this book and tabulated the calibration strategies in the respective chapters. At the end of each chapter, I have added appendix materials, including the original and calibrated dataset, and the calibration codes for all my variables. I have also uploaded these codes to Harvard Dataverse, https://doi.org/10.7910/DVN/F0PT9C.

DOI: 10.1201/9781003384595-4

TABLE 4.1   Installing and Running R Packages

```
Installing R Packages:
install.packages("readxl")
install.packages("dplyr")
install.packages("ggplot2")
install.packages("QCA")
install.packages("SetMethods")
install.packages("readr")
install.packages("rmarkdown")
install.packages("tinytex")
install.packages("knitr")

Running R Packages:
library("readxl")
library("dplyr")
library("ggplot2")
library("QCA")
library("SetMethods")
library("readr")
library("rmarkdown")
library("tinytex")
library("knitr")
```

TABLE 4.2   Downloading the Dataset

```
data<-read.csv("Panel_data.csv")
str(data)
data%>%slice_head(n=5)
summary(data)
```

Once we have installed and run the packages, we can then download the dataset, as shown above in Table 4.2. Since we are working in RStudio, we need to check and set the working directory with *getwd()* and *setwd(),* first.

I have applied the pnames() function to add row names for each country_year since the data for Chapter 5 is in long format.

CALIBRATING CRISP SETS

In crisp-set QCA, there are two main qualitative states, 0 (non-membership) and 1 (membership). To calibrate a crisp-set QCA, we need to first define the set and then set a threshold to establish full membership.

For example, I can apply "Presence of microfinance institutions (MFIs)", as a necessary condition. I will first define the set as "Countries with presence of MFIs." Since this is a dichotomous categorical variable, I have calibrated it as crisp-set, with zero indicating "countries with absence of microfinance institutions" and one indicating "countries with presence of microfinance institutions."

Before calibrating this variable, into a crisp-set, we need to clean the data, summarize, and visualize it, as seen in Table 4.3, and Table 4.4. Due to limited space, I have included my data cleaning process as a part of my calibration code chunk.

TABLE 4.3 Codes for Calibrating "Set of Countries with Presence of Microfinance Institutions"

```
str(data$PresenceofMFIs)
Xplot(data$PresenceofMFIs, jitter=TRUE)
data$MFISR<-calibrate(data$PresenceofMFIs, type="crisp",
 thresholds=1)
skew.check(data$MFISR)
print(data[1:5,c("Observations", "PresenceofMFIs",
 "MFISR")])
```

PRESENCE OF MFIS

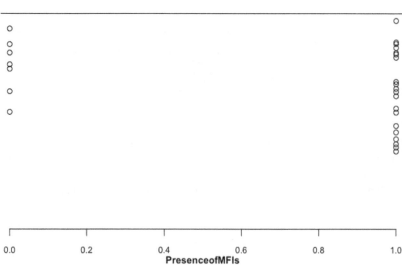

After calibrating the data, we can apply the skew.check() function to check for skewness and the plot() function, to plot the original data (as x-axis) with the calibrated data (as y-axis). My calibrated data does show a high percentage of skewness since most of my observations had microfinance institutions present. As discussed by Oana, Schneider, and Thomann (2021, 47–50), a high skewness can lead to trivial necessary and sufficient conditions, and researchers should address skewness, by re-calibrating the variable. Though I have not addressed skewness in this book, readers are encouraged to refer to Oana, Schneider, and Thomann (2021), for further guidelines. I have applied the plot() function below, for my fuzzy-set calibrations.

CALIBRATING FUZZY-SETS

There are three main ways to create fuzzy-sets in QCA:

1. Applying a theory-guided approach
2. Applying Ragin's Direct Calibration approach
3. Applying Ragin's Indirect Calibration approach

I demonstrate each of these approaches below.

APPLYING THEORY-GUIDED CALIBRATION TO CREATE FUZZY-SETS

In theory-guided calibration, we can apply existing theory, case study knowledge, or existing external criteria, to create a four-value or five-value fuzzy-set. This is known as "Theory-Guided Calibration" (Oana, Schneider, and Thomann 2021, 38–41). To do this, we need to first establish qualitative thresholds.

As an example, I have calibrated my variable "Countries with High European Union membership" into a four-value fuzzy-set, codes in Table 4.4. I have calibrated the variable based on external existing criteria of EU membership stages: non-members, potential candidates, candidate countries, and

TABLE 4.4 Theory-Guided Calibration, "Set of Countries with High European Union Membership"

```
data$THEUM<-NA
data$THEUM[data$EUM >=3]<-1
data$THEUM[data$EUM<3 & data$EUM>=2]<-0.67
data$THEUM[data$EUM<2 & data$EUM>=1]<-0.33
data$THEUM[data$EUM<1]<-0
data$THEUM
Or
data$EUMTHR<-recode(data$EUM, cuts="0, 1, 2", values="0,
 0.33, 0.67, 1")
print(data[1:5,c("EUM","THEUM", "EUMTHR")])
skew.check(data$THEUM)
skew.check(data$EUMTHR)
```

full members, and created the qualitative thresholds of 0 (for non-members), 0.33 (for potential candidates), 0.67 (for candidate countries) and 1 (for full members). We can also apply the recode() function to directly calibrate the condition into a set. I have added both these codes above.

APPLYING RAGIN'S DIRECT CALIBRATION APPROACH

Ragin's Direct Calibration approach is very similar to the theory-guided calibration approach discussed above, but this approach can be used to create only a three-value fuzzy-set, based on the qualitative thresholds are 0 (full-set exclusion), 0.5 (crossover point), and 1 (full-set inclusion), and should be based on existing theory, case study knowledge, or existing external criteria (Oana, Schneider, and Thomann 2021, 42–45).

For example, in my causal analysis model, I have measured the extent of political and civil liberty, through the Freedom House Report. This variable ranges from 1.0 to 7.0, with 1.0 to 2.5 categorized as "Free," 3.0 to 5.0 categorized as "Partly Free," and 5.5 to 7.0 categorized as "Not Free" (Freedom House Report, Political Rights and Civil Liberties).

To calibrate this condition, I have first defined it as a "set of countries with high political and civil liberty". I then applied Ragin's Direct Calibration and established the qualitative thresholds of (1.9) as full-set inclusion, (2.6) as crossover point, and (5.1) as full-set exclusion. I have tabulated the codes in Table 4.5 and visualized them in Figures 4.2–4.5.

TABLE 4.5 Codes for Applying Ragin's Direct Calibration Method

```
ggplot(data,aes(x=YEAR,y=PCL,
 group=COUNTRY))+geom_line()+facet_wrap(.~COUNTRY)+theme_
 light()
print(data$PCL)
sort(data$PCL)
Xplot(data$PCL, jitter=TRUE)
hist(data$PCL)
data$DPCL<-calibrate(data$PCL, type="fuzzy",
thresholds="i=1.9,c=2.6, e=5.1", logistic = TRUE)
plot(data$PCL,data$DPCL, xlab="Raw Score", ylab="Calibrated
Score", abline (h=0.5,v=2.6))
skew.check(data$DPCL)
print(data[1:5,c("PCL","DPCL")])
```

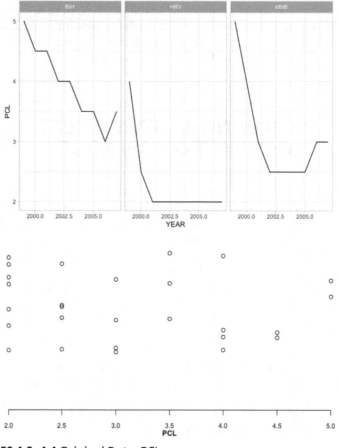

FIGURES 4.2–4.4 Original Data, PCL

Histogram of data$PCL

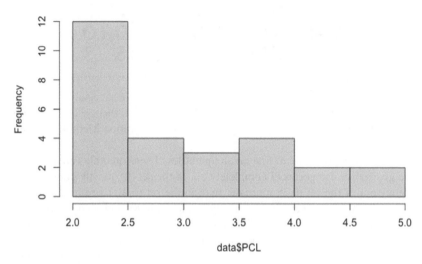

FIGURE 4.2-4.4 (*Continued*)

Calibration of PCL

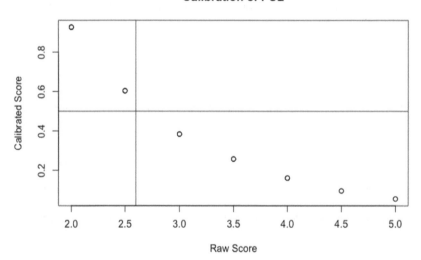

FIGURE 4.5 Calibrated Data, DPCL

APPLYING RAGIN'S INDIRECT CALIBRATION METHOD TO CREATE FUZZY-SETS

If we have an interval or ratio-scale variable, but we cannot establish qualitative thresholds based on external social science criteria, theory, or case study knowledge, then we can apply Ragin's Indirect Calibration Method to create fuzzy-sets.

To do this, we need to first create quantiles. These quantiles are based on the qualitative categories of completely out of the set (0), mostly but not completely out of the set (0.2), more out than in the set (0.4), more in than out of the set (0.6), mostly but not completely in the set (0.8), and completely in the set (1) (Bhattacharya 2020, 2023b; Dusa 2022). We can then apply different types of regression models – such as beta or binomial regression, if non-linear data – to transform the data into set-membership scores ranging from 0 to 1 (Bhattacharya 2023b; Dusa 2022, 106–108).

I have applied one of my outcome variables, "economic participation of women as female own-account workers,", OWN, to demonstrate this. I have tabulated the codes, in Table 4.6, and visualized the calibrated data in Figures 4.6–4.8.

TABLE 4.6   Codes for Ragin's Indirect Calibration

```
Xplot(data$OWN, jitter=TRUE, at=pretty(data$OWN))
sort(data$OWN)
quantfown <- quantile(data$OWN,c(0.2,0.4,0.5,0.6,0.8))
quantfown
data$OWNB<-NA
data$OWNB[data$OWN<= quantfown [1]]<-0
data$OWNB[data$OWN > quantfown [1] & data$OWN <= quantfown
 [2]]<-0.2
data$OWNB[data$OWN > quantfown [2] & data$OWN <= quantfown
 [3]]<-0.4
data$OWNB[data$OWN > quantfown [3] & data$OWN <= quantfown
 [4]]<-0.6
data$OWNB[data$OWN > quantfown [4] & data$OWN <= quantfown
 [5]]<-0.8
data$OWNB[data$OWN > quantfown [5]]<-1
data$OWNbinom <- indirectCalibration(data$OWN, data$OWNB,
 binom=TRUE)
```

```
data$OWNbeta <- indirectCalibration(data$OWN, data$OWNB,
 binom=FALSE)
cor(data$OWNbinom,data$OWNbeta)
plot(data$OWN,data$OWNbinom)
plot(data$OWN,data$OWNbeta)
skew.check(data$OWNbinom)
skew.check(data$OWNbeta)
```

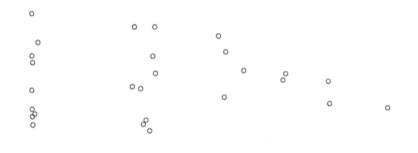

FIGURE 4.6 Original Data, OWN

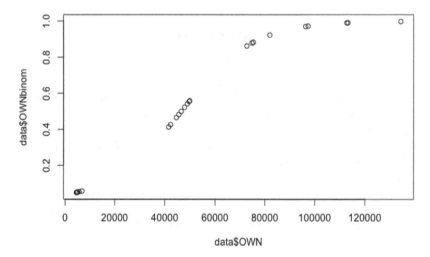

FIGURE 4.7 Calibrated Data, Ownbinom

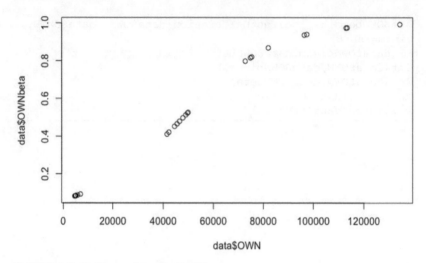

FIGURE 4.8: Calibrated Data, OWNbeta

The skewness check shows that calibration through beta regression (44.44%, Figure 4.7) has a lower extent of skewness as compared to the extent of skewness from binomial regression, though they both have a high percentage of skewness (48.15%, Figure 4.8). Hence, I have selected OWNbeta as my calibrated variable.

CONCLUSION

Once we have calibrated all the conditions and outcomes as sets, we can then start testing for necessary and sufficient conditions. In the next few chapters, chapters 5–8, I will demonstrate each of the 4 panel data QCA models, as mentioned in chapter 1. I have tabulated my calibration strategy for all my conditions and outcome variables, for this chapter and Chapter 5, in Table 4.7 and tabulated the remaining calibration strategies, in their individual chapters.

Along with this, I have added my original data, calibration codes for all my variables, and my calibrated dataset, in this chapter's appendix, and also uploaded these codes and datasets on Harvard Dataverse, https://doi.org/10.7910/DVN/F0PT9C.

TABLE 4.7 Causal Analysis Model (Conditions and Outcome Variables)

CONCEPT	DIMENSION/ CHARACTERISTIC	INDICATOR/ VARIABLE	ABBREVIATIONS	TYPE OF SET	TYPE OF CALIBRATION	SET LABEL
Presence of Microfinance	Number of Microfinance Institutions	Presence of MFIS Source: World Bank| Databank| MIX Market	MFISR	Crisp	Crisp-set	Presence of Microfinance
Access to Political Freedom	Extent of political and civil rights Extent of democracy within the country	Political Rights & Civil Liberty Scale Source: Freedom Rating Scale, Freedom House Reports Democracy Scale, Source: Polity IV Dataset	DPCL DEMC	Fuzzy	Ragin's Direct Calibration Method (thresholds: i=1.9, c=2.6, e=5.1) Ragin's Direct Calibration Method (thresholds: e=67, c=5.3, i=7)	Set of countries with high political and civil liberty –––––––––– Set of highly democratic countries
Access to Economic Facilities	Economic growth -> increases opportunities available to individuals for production, consumption, and exchange	Gross Domestic Product Source: World Bank	GDPR	Fuzzy	Ragin's Direct Calibration Method (thresholds: e= 980000000, c=30000000000, i=50000000000)	Set of countries with High GDP

(Continued)

TABLE 4.7 (Continued)

CONCEPT	DIMENSION/ CHARACTERISTIC	INDICATOR/ VARIABLE	ABBREVIATIONS	TYPE OF SET	TYPE OF CALIBRATION	SET LABEL
Access to Social Opportunities	Access to basic education for females	% of literate females between the ages of 15 to 24 years. Source: World Bank	LTRCYR	Fuzzy	Ragin's Direct Calibration Method (thresholds: e=98.70, c=99.70, i=99.84)	Set of countries with high percentage of literate females
	Access to health care for females	Life expectancy at birth for females. Measured in years. Source: World Bank	LIFEXFR		Ragin's Direct Calibration Method (thresholds: e=76.00, c=77.85, i=79.00)	Set of countries with high life expectancy for females
Access to transparency guarantees	Society operates on the basic presumption of trust. Need for openness, guarantees of disclosure, and lucidity	Control of corruption Scale Source: World Governance Indicators| World Bank	FCORR	Fuzzy	Ragin's Direct Calibration Method (thresholds: e=33.00, c=55.00, i=71.00)	Set of countries with High Control of Corruption
Access to protective security	Effectiveness of public institutions in ensuring access to basic facilities like sanitation, health, and unemployment benefits	Government Effectiveness Scale Source: World Governance Indicators| World Bank	NEFFECTR	Fuzzy	Ragin's Direct Calibration Method (thresholds: e=10.00, c=55.00, i=69.00)	Set of countries with High Government Effectiveness

Condition	Definition	Details / Source	Code	Type	Calibration	Set
European Union membership	Are the countries full members of the EU?	Joining the EU: 27 members of the EU. Source: European Union database (Case Study Condition)	THEUM/EUMTHR	Fuzzy	Theory-Guided Calibration	Set of Countries with High European Union membership
Signed the 18 International Human Rights Treaties	Members of Human Rights Treaties to eliminate all forms of racial and gender equality	Ratification of 18 International Human Rights Treaties, Source: UN Human Rights Office of the High Commissioner	HRTR	Fuzzy	Ragin's Direct Calibration Method (thresholds: e=0, c=10.5, i=14)	Set of countries with high number of International Human Rights Treaties
Macro-condition: Economic participation of women	Outcome 1: Economic participation of women as own-account workers (*Self-employed, do not have any persons working for them*)	Number of women employed belonging to the age group 15 – 65 years. Source: Employment by Sex and Status in Employment (KILM 3); ILO Modeled Estimates – Annual, calculated in thousands	HNWEB: fuzzyand(OWNR, FAMR, EMPER, EMPRR)	Fuzzy	Ragin's Direct Calibration Method (thresholds: e= 4000, c=77000, i=110000)	Set of countries with high number of female own-account workers
					Ragin's Direct Calibration Method (thresholds: e= 1700, c=43000, i= 60000)	
	Outcome 2: participation of women as female family workers (*Self-employed, has family members working for them*)				Ragin's Direct Calibration Method (thresholds: e= 55000, c=300000, i= 610000)	Set of countries with high number of female family workers

TABLE 4.7 (Continued)

CONCEPT	DIMENSION/ CHARACTERISTIC	INDICATOR/ VARIABLE	ABBREVIATIONS	TYPE OF SET	TYPE OF CALIBRATION	SET LABEL
	---------- Outcome 3: Economic participation of women as female employees *(Holds paid employment jobs with remuneration & contract)*				Ragin's Direct Calibration Method (thresholds: e= 3000, c=15000, i=25000)	Set of countries with high number of female employees
	---------- Outcome 4: Economic participation of women as female employers *(Self-employed, but has employees working for them with paid employment contract)*					---------- Set of countries with high number of female employers

Source: Table built based on concepts, as defined by Goertz (2006)

Appendix

- **Original dataset: <Panel_data.csv>**

OBSERVATIONS	MFIRSD	GDPE	DEMD	PCLE	LTRCYFE
1 Bosnia and Herzegovina	0.272598	0.968481	0.95	0.272598	0.951162
2 Croatia	0.05	0.48531	0	0.05	0.246395
3 Montenegro	0.95	0.04849	0.05	0.965069	0.046253

OBSERVATIONS	LIFEXFE	NCORD	NEFFECTE	FRPRD	HRTRD
1 Bosnia and Herzegovina	0.104611	0.044894	0.047402	0.962041	0.064435
2 Croatia	0.367923	0.950838	0.313009	0.05	0.379719
3 Montenegro	0.346544	0.190918	0.951209	0.862981	0.573087

OBSERVATIONS	OWNE	FAMD	EMPEF	EMPRD	HNWEB
1 Bosnia and Herzegovina	0.046634	0.984155	0.759114	0.957926	0.046634
2 Croatia	0.3345	0.049057	0.047617	0.933873	0.047617
3 Montenegro	0.95878	0.302962	0.934028	0.049279	0.049279

- **Calibration codes for all the remaining variables**

Calibrating "set of countries with high number of international human rights treaties"

```
hist(data$INTHRT)
str(data$INTHRT)
sort(data$INTHRT)
Xplot(data$INTHRT, jitter=TRUE)
data$HRTR<-calibrate(data$INTHRT, type="fuzzy",
thresholds="e=0, c=10.5, i=14", logistic=TRUE)
skew.check(data$HRTR)
print(data[1:5,c("INTHRT","HRTR")])
plot(data$INTHRT, data$HRTR, xlab="Raw Score",
ylab="Calibrated Score", abline(h=0.5, v=10.5))
```

Calibrating "set of countries with high GDP"

```
ggplot(data,
aes(x=YEAR,y=GDP,group=COUNTRY))+geom _
line(aes(col=COUNTRY)) + theme _ light()
print(data$GDP)
sort(data$GDP)
hist(data$GDP)
Xplot(data$GDP, jitter=TRUE)
data$GDPR<-calibrate(data$GDP, thresholds="e=980000000,
c=30000000000, i=50000000000", logistic=TRUE)
skew.check(data$GDPR)
print(data[1:5,c("GDP","GDPR")])
plot(data$GDP, data$GDPR, xlab="Raw Score",
ylab="Calibrated Score", abline(h=0.5, v=30000000000))
```

Calibrating "set of highly democratic countries"

```
ggplot(data, aes(x=YEAR,y=DEM,group=COUNTRY))+geom _
line()+facet _ wrap(.~COUNTRY) + theme _ light()
print(data$DEM)
str(data$DEM)
hist(data$DEM)
```

```
xplot(data$DEM, jitter=TRUE)
data$DEMC<-calibrate(data$DEM, thresholds="e=-
67,c=5.3,i=7", logistic=TRUE)
skew.check(data$DEMC)
print(data[1:5,c("DEM","DEMC")])
plot(data$DEM,data$DEMC, xlab="Raw Score",
ylab="Calibrated Score", abline (h=0.5,v=5.3))
```

Calibrating "set of countries with high percentage of literate females"

```
print(data[,c("Observations","LTRCYF")])
sort(data$LTRCYF)
Xplot(data$LTRCYF, jitter=TRUE)
hist(data$LTRCYF)
data$LTRCYR<-calibrate(data$LTRCYF, type="fuzzy", thres
holds="e=98.70,c=99.70,i=99.84", logistic = TRUE)
skew.check(data$LTRCYR)
print(data[1:5,c("Observations","LTRCYF", "LTRCYR")])
```

Calibrating "set of countries with high life expectancy for females"

```
sort(data$LIFEXF)
Xplot(data$LIFEXF, jitter=TRUE)
hist(data$LIFEXF)
data$LIFEXFR<-calibrate(data$LIFEXF, type="fuzzy", thre
sholds="e=76.0,c=77.85,i=79.00", logistic = TRUE)
skew.check(data$LIFEXFR)
print(data[1:5,c("Observations","LIFEXF", "LIFEXFR")])
plot(data$LIFEXF,data$LIFEXFR, xlab="Raw Score",
ylab="Calibrated Score", abline (h=0.5,v=77.85))
```

Calibrating "set of countries with high government effectiveness"

```
print(data[,c("Observations","FEFFECT")])
data$NEFFECT<- as.numeric(data$FEFFECT)
data$NEFFECT[which(data$COUNTRY=="BiH" & data$YEAR
%in% c(1999))]<-11.398960000000001
```

```
data$NEFFECT[which(data$COUNTRY=="HRV" & data$YEAR
%in% c(1999))]<-59.58549

data$NEFFECT[which(data$COUNTRY=="MNE" & data$YEAR
%in% c(1999,2000,2001,2002,2003,2004))]<-0

data$NEFFECT[which(data$COUNTRY=="BiH" & data$YEAR
%in% c(2001))]<-17.948720000000002

data$NEFFECT[which(data$COUNTRY=="HRV" & data$YEAR
%in% c(2001))]<-66.666659999999993

print(data$NEFFECT)

sort(data$NEFFECT)

Xplot(data$NEFFECT, jitter=TRUE)

hist(data$NEFFECT)

data$NEFFECTR<-calibrate(data$NEFFECT, type="fuzzy",
thresholds="e=10.0, c=55.0, i=69.0", logistic = TRUE)

skew.check(data$NEFFECTR)

plot(data$NEFFECT,data$NEFFECTR, xlab="Raw Score",
ylab="Calibrated Score", abline(h=0.5,v=55.0))

print(data[1:5,c("Observations", "NEFFECT", "NEFFECTR")])
```

Calibrating "set of countries with high female representation"

```
print(data[,c("Observations","FPARL")])

data$FPARL[which(data$COUNTRY=="BiH" & data$YEAR %in%
c(1999, 2000))]<-0

data$FPARL[which(data$COUNTRY=="BiH" & data$YEAR %in%
c(2002))]<-7.142

data$FPARL[which(data$COUNTRY=="HRV" & data$YEAR %in%
c(2003))]<-20.520

data$FPARL[which(data$COUNTRY=="MNE" & data$YEAR %in%
c(1999, 2000, 2001, 2002, 2003, 2004, 2005))]<-0

Xplot(data$FPARL, jitter=TRUE, at=pretty(data$FPARL))

hist(data$FPARL)

sort(data$FPARL)

data$FRPR<-calibrate(data$FPARL, type="fuzzy",
thresholds="e=7.00, c=17.00, i=21.00", logistic = TRUE)

skew.check(data$FRPR)

print(data[1:5,c("Observations","FPARL", "FRPR")])

plot(data$FPARL,data$FRPR, xlab="Raw Score",
ylab="Calibrated Score", abline(h=0.5,v=17.00))
```

Calibrating "set of countries with high number of female own-account workers"

```
Xplot(data$OWN, jitter=TRUE, at=pretty(data$OWN))
sort(data$OWN)
hist(data$OWN)
data$OWNR<-calibrate(data$OWN, type="fuzzy",
thresholds="e=4000, c=77000, i=110000", logistic = TRUE)
skew.check(data$OWNR)
print(data[1:5,c("Observations","OWN", "OWNR")])
plot(data$OWN,data$OWNR,xlab="Raw Score",
ylab="Calibrated Score", abline(h=0.5,v=77000))
```

Calibrating "set of countries with high number of female family workers"

```
Xplot(data$FAM, jitter=TRUE, at=pretty(data$FAM))
sort(data$FAM)
hist(data$FAM)
data$FAMR<-calibrate(data$FAM, type="fuzzy",
thresholds="e=1700, c=43000, i=60000", logistic = TRUE)
skew.check(data$FAMR)
print(data[1:5,c("Observations","FAM", "FAMR")])
plot(data$FAM,data$FAMR, xlab="Raw Score",
ylab="Calibrated Score", abline(h=0.5,v=43000))
```

Calibrating "set of countries with high number of female employees"

```
Xplot(data$EMPE, jitter=TRUE, at=pretty(data$EMPE))
sort(data$EMPE)
hist(data$EMPE)
data$EMPER<-calibrate(data$EMPE, type="fuzzy",
thresholds="e=55000, c=300000, i=610000", logistic = TRUE)
skew.check(data$EMPER)
print(data[1:5,c("Observations","EMPE", "EMPER")])
plot(data$EMPE,data$EMPER, xlab="Raw Score",
ylab="Calibrated Score", abline(h=0.5,v=300000))
```

Calibrating "set of countries with high number of female employers"

```
Xplot(data$EMPR,jitter=TRUE, at=pretty(data$EMPR))
sort(data$EMPR)
hist(data$EMPR)
data$EMPRR<-calibrate(data$EMPR, type="fuzzy",
thresholds="e=3000, c=15000, i=25000", logistic = TRUE)
skew.check(data$EMPRR)
print(data[1:5,c("Observations","EMPR", "EMPRR")])
plot(data$EMPR,data$EMPRR, xlab="Raw Score",
ylab="Calibrated Score", abline(h=0.5,v=15000))
```

Calibrating "set of countries with high control of corruption"

```
print(data[,c("Observations","FCOR")])
data$FCOR[which(data$COUNTRY=="BiH" & data$YEAR %in%
c(1999))]<-42.783499999999997
data$FCOR[which(data$COUNTRY=="HRV" & data$YEAR %in%
c(1999))]<-34.536079999999998
data$FCOR[which(data$COUNTRY=="MNE" & data$YEAR %in%
c(1999))]<-71.134020000000007
data$FCOR[which(data$COUNTRY=="BiH" & data$YEAR %in%
c(2001))]<-33.502540000000003
data$FCOR[which(data$COUNTRY=="HRV" & data$YEAR %in%
c(2001))]<-56.345179999999999
data$FCOR[which(data$COUNTRY=="MNE" & data$YEAR %in%
c(2001))]<-53.299489999999999
ggplot(data, aes(x=YEAR,y=FCOR, group=COUNTRY)) +
geom _ line(aes(col=COUNTRY)) + theme _ light()
Xplot(data$FCOR,jitter=TRUE,at=pretty(data$FCOR))
hist(data$FCOR)
sort(data$FCOR)
data$FCORR<-calibrate(data$FCOR, type="fuzzy",
thresholds="e=33.0, c=55.0, i=71.0", logistic = TRUE)
skew.check(data$FCORR)
print(data[1:5,c("Observations","FCOR", "FCORR")])
plot(data$FCOR,data$FCORR, xlab="Raw Score",
ylab="Calibrated Score", abline(h=0.5,v=55.0))
```

Creating a macro-condition, "set of countries with high economic participation of women at the national level", as the outcome variable:

```
skew.check(data$HNWE)
data$HNWEB <- with(data,fuzzyand(OWNR, FAMR, EMPER,
EMPRR))
skew.check(data$HNWEB)
print(data[1:5,c("Observations", "OWNR", "FAMR", "EMPER",
"EMPRR", "HNWEB")])
```

Creating a new dataset:

```
PANEL<-data %>% select (Observations, MFISR, GDPR,
DEMC, DPCL, LTRCYR, LIFEXFR, FCORR, NEFFECTR, FRPR,
HRTR, OWNR, FAMR, EMPER, EMPRR, HNWEB, COUNTRY, YEAR)
PANEL<-write.csv("PANEL.csv")
```

Creating a latex file, optional:

```
knit("Chapter4 _ withAppendix.Rmd")
pandoc _ convert("Chapter4 _ withAppendix.Rmd", to
="latex", output="chapter4.tex")
```

- *Insert the calibrated panel dataset here, PANEL.csv, here*

	OBSERVATIONS	MFISR	GDPR	DEMC	DPCL	LTRCYR
BIH99	Bosnia and Herzegovina_1999	1	0.071197	0.05197	0.0559	0.055197
HRV99	Croatia_1999	1	0.338586	0.446248	0.16126	0.430089
MNE99	Montenegro_1999	0	0.23577	0.446248	0.0559	0.109791
BIH00	Bosnia and Herzegovina_2000	1	0.0769	0.05197	0.096409	0.953504

(Continued)

	OBSERVATIONS	MFISR	GDPR	DEMC	DPCL	LTRCYR
HRV00	Croatia_2000	1	0.300427	0.446248	0.603635	0.430089
MNE00	Montenegro_2000	0	0.050021	0.95	0.16126	0.109791
BIH01	Bosnia and Herzegovina_2001	1	0.078669	0.05197	0.096409	0.953504
HRV01	Croatia_2001	1	0.33359	0.446248	0.925794	0.465612
MNE01	Montenegro_2001	0	0.050874	0.95	0.384353	0.109791
BIH02	Bosnia and Herzegovina_2002	1	0.085565	0.05197	0.16126	0.953504
HRV02	Croatia_2002	1	0.419924	0.446248	0.925794	0.465612
MNE02	Montenegro_2002	0	0.051489	0.95	0.603635	0.109791
BIH03	Bosnia and Herzegovina_2003	1	0.100232	0.05197	0.16126	0.953504
HRV03	Croatia_2003	1	0.665959	0.446248	0.925794	0.465612
MNE03	Montenegro_2003	0	0.053626	0.770726	0.603635	0.269071
BIH04	Bosnia and Herzegovina_2004	1	0.122523	0.05197	0.25731	0.953504
HRV04	Croatia_2004	1	0.846441	0.446248	0.925794	0.465612
MNE04	Montenegro_2004	0	0.05554	0.770726	0.603635	0.269071
BIH05	Bosnia and Herzegovina_2005	1	0.129525	0.05197	0.25731	0.953504
HRV05	Croatia_2005	1	0.905959	0.446248	0.925794	0.465612
MNE05	Montenegro_2005	0	0.056527	0.770726	0.603635	0.269071
BIH06	Bosnia and Herzegovina_2006	1	0.149491	0.05197	0.384353	0.953504
HRV06	Croatia_2006	1	0.953087	0.446248	0.925794	0.465612
MNE06	Montenegro_2006	1	0.059095	0.998355	0.384353	0.269071
BIH07	Bosnia and Herzegovina_2007	1	0.191094	0.05197	0.25731	0.953504
HRV07	Croatia_2007	1	0.988383	0.446248	0.925794	0.465612
MNE07	Montenegro_2007	1	0.064742	0.998355	0.384353	0.269071

	OBSERVATIONS	LIFEXFR	FCORR	NEFFECTR	FRPR
BIH99	Bosnia and Herzegovina_1999	0.117147	0.163142	0.054531	0.006656
HRV99	Croatia_1999	0.11214	0.060719	0.724003	0.064198
MNE99	Montenegro_1999	0.059002	0.951159	0.026629	0.006656
BIH00	Bosnia and Herzegovina_2000	0.176209	0.053293	0.081335	0.006656
HRV00	Croatia_2000	0.134452	0.561573	0.920831	0.929326
MNE00	Montenegro_2000	0.052713	0.443346	0.026629	0.006656
BIH01	Bosnia and Herzegovina_2001	0.240649	0.053293	0.081335	0.051417
HRV01	Croatia_2001	0.694092	0.561573	0.920831	0.929326
MNE01	Montenegro_2001	0.05169	0.443346	0.026629	0.006656
BIH02	Bosnia and Herzegovina_2002	0.305159	0.165823	0.063127	0.052024
HRV02	Croatia_2002	0.803483	0.830521	0.906773	0.929326
MNE02	Montenegro_2002	0.053112	0.45115	0.026629	0.006656
BIH03	Bosnia and Herzegovina_2003	0.367182	0.241888	0.109438	0.477931
HRV03	Croatia_2003	0.725706	0.843205	0.930654	0.930287
MNE03	Montenegro_2003	0.055398	0.147958	0.026629	0.006656
BIH04	Bosnia and Herzegovina_2004	0.424582	0.205103	0.159118	0.477931
HRV04	Croatia_2004	0.95888	0.811342	0.949654	0.96952
MNE04	Montenegro_2004	0.05865	0.093747	0.026629	0.006656
BIH05	Bosnia and Herzegovina_2005	0.477335	0.360946	0.130243	0.477931
HRV05	Croatia_2005	0.924782	0.657197	0.928029	0.96952
MNE05	Montenegro_2005	0.064725	0.227211	0.874156	0.006656
BIH06	Bosnia and Herzegovina_2006	0.544687	0.194663	0.183596	0.311096
HRV06	Croatia_2006	0.97789	0.696433	0.957035	0.96952
MNE06	Montenegro_2006	0.077181	0.11841	0.46217	0.077749
BIH07	Bosnia and Herzegovina_2007	0.623492	0.137211	0.094088	0.182177
HRV07	Croatia_2007	0.973	0.665493	0.949317	0.946385
MNE07	Montenegro_2007	0.1021	0.257499	0.380842	0.149667

(Continued)

	OBSERVATIONS	HRTR	OWNR	FAMR	EMPER
BIH99	Bosnia and Herzegovina_1999	0.331578	0.250858	0.942073	0.331278
HRV99	Croatia_1999	0.331578	0.609782	0.988983	0.931322
MNE99	Montenegro_1999	0.05	0.051995	0.051164	0.050363
BIH00	Bosnia and Herzegovina_2000	0.331578	0.243793	0.878298	0.337492
HRV00	Croatia_2000	0.331578	0.48361	0.954056	0.922859
MNE00	Montenegro_2000	0.05	0.051745	0.051171	0.050345
BIH01	Bosnia and Herzegovina_2001	0.396365	0.235474	0.787066	0.339868
HRV01	Croatia_2001	0.396365	0.478354	0.90098	0.924151
MNE01	Montenegro_2001	0.05	0.051595	0.050943	0.050466
BIH02	Bosnia and Herzegovina_2002	0.603635	0.226039	0.67868	0.338597
HRV02	Croatia_2002	0.603635	0.457359	0.876828	0.919359
MNE02	Montenegro_2002	0.05	0.051411	0.050747	0.050537
BIH03	Bosnia and Herzegovina_2003	0.779353	0.219288	0.555498	0.341118
HRV03	Croatia_2003	0.603635	0.860247	0.71537	0.921744
MNE03	Montenegro_2003	0.05	0.05126	0.050562	0.050743
BIH04	Bosnia and Herzegovina_2004	0.779353	0.21263	0.476224	0.342757
HRV04	Croatia_2004	0.603635	0.962372	0.438378	0.93354
MNE04	Montenegro_2004	0.05	0.051162	0.050377	0.051073
BIH05	Bosnia and Herzegovina_2005	0.779353	0.197839	0.411711	0.326865
HRV05	Croatia_2005	0.779353	0.994085	0.29738	0.936181
MNE05	Montenegro_2005	0.05	0.051101	0.05017	0.051459
BIH06	Bosnia and Herzegovina_2006	0.779353	0.192943	0.375118	0.329162
HRV06	Croatia_2006	0.779353	0.96081	0.258347	0.952096
MNE06	Montenegro_2006	0.465004	0.053386	0.05106	0.055541
BIH07	Bosnia and Herzegovina_2007	0.779353	0.249006	0.282948	0.355322
HRV07	Croatia_2007	0.95	0.84924	0.268233	0.958145
MNE07	Montenegro_2007	0.603635	0.055775	0.051735	0.059092

	OBSERVATIONS	EMPRR	HNWEB	COUNTRY	YEAR
BIH99	Bosnia and Herzegovina_1999	0.260369	0.250858	BiH	1999
HRV99	Croatia_1999	0.938728	0.609782	HRV	1999
MNE99	Montenegro_1999	0.053057	0.050363	MNE	1999
BIH00	Bosnia and Herzegovina_2000	0.261837	0.243793	BiH	2000
HRV00	Croatia_2000	0.910634	0.48361	HRV	2000
MNE00	Montenegro_2000	0.053379	0.050345	MNE	2000
BIH01	Bosnia and Herzegovina_2001	0.260937	0.235474	BiH	2001
HRV01	Croatia_2001	0.701135	0.478354	HRV	2001
MNE01	Montenegro_2001	0.054002	0.050466	MNE	2001
BIH02	Bosnia and Herzegovina_2002	0.257544	0.226039	BiH	2002
HRV02	Croatia_2002	0.965301	0.457359	HRV	2002
MNE02	Montenegro_2002	0.054505	0.050537	MNE	2002
BIH03	Bosnia and Herzegovina_2003	0.257357	0.219288	BiH	2003
HRV03	Croatia_2003	0.819614	0.71537	HRV	2003
MNE03	Montenegro_2003	0.055141	0.050562	MNE	2003
BIH04	Bosnia and Herzegovina_2004	0.256607	0.21263	BiH	2004
HRV04	Croatia_2004	0.856303	0.438378	HRV	2004
MNE04	Montenegro_2004	0.055513	0.050377	MNE	2004
BIH05	Bosnia and Herzegovina_2005	0.239988	0.197839	BiH	2005
HRV05	Croatia_2005	0.885552	0.29738	HRV	2005
MNE05	Montenegro_2005	0.055758	0.05017	MNE	2005
BIH06	Bosnia and Herzegovina_2006	0.240256	0.192943	BiH	2006
HRV06	Croatia_2006	0.947216	0.258347	HRV	2006
MNE06	Montenegro_2006	0.068607	0.05106	MNE	2006
BIH07	Bosnia and Herzegovina_2007	0.248273	0.248273	BiH	2007
HRV07	Croatia_2007	0.91225	0.268233	HRV	2007
MNE07	Montenegro_2007	0.084457	0.051735	MNE	2007

Cluster QCA

5

Chapter outline:

- Introduction
- Assumptions
- Data calibration
- Testing for necessary conditions
- Interpreting results from the testing of necessary conditions
- Testing for sufficient conditions
- Interpreting the enhanced intermediate solution
- Applying the cluster() function
- Interpreting results from Step 5
- Visualizing results from testing of necessary and sufficient conditions
- A few other interesting features
- Strengths of Cluster QCA
- Weaknesses
- Appendix

INTRODUCTION

In this chapter, I will demonstrate Garcia-Castro and Ariño's (2016) panel data QCA model. I have termed this approach as *Cluster QCA*.

Garcia-Castro and Ariño, developed this model, in their article "A General Approach to Panel Data Set-Theoretic Research" (2016). To demonstrate this approach, I first discuss the assumptions of this model. I then discuss how to interpret consistency and coverage scores, test necessary and sufficient conditions, and interpret the data analysis results. I also discuss the cluster() function, and how to interpret the cluster results. I conclude this chapter by demonstrating how to visualize data analysis results, the importance of this approach, and the most important drawbacks of this model.

DOI: 10.1201/9781003384595-5

ASSUMPTIONS

The assumptions of this panel data QCA model are mostly similar to cross-sectional QCA, with one additional assumption: we need to interpret consistency and coverage cross-sectionally as well as over-time (Garcia-Castro and Ariño 2016). The across-time consistency and coverage act as a robustness check for cross-sectional consistency and coverage, and help researchers understand whether the impact of conditions varies over the time period and cases studied (Garcia-Castro and Ariño 2016, 63). Hence, there are three main types of consistency scores – within consistency, between consistency, and pooled consistency; and three main types of coverage scores – within coverage, between coverage, and pooled coverage, tabulated in Table 5.1 (Garcia-Castro and Ariño 2016; Furnari 2018; Dusa 2022; Oana, Schneider, and Thomann 2021; Bhattacharya 2023a).

The *within consistency* (WICONS) can be defined as the longitudinal consistency of the set-subset relation, for each case in the panel over time, by holding the time as constant (Garcia-Castro and Ariño 2016). The *between consistency* (BECONS) can be defined as the cross-sectional consistency of the set-subset relation for each year in the panel, by holding the case as constant (Garcia-Castro and Ariño 2016). The *pooled consistency* (POCON) refers to the overall consistency, when the country or time is not taken into account (Garcia-Castro and Ariño 2016). In a panel data QCA model, the number of *WICONS* depends on the number of cases/countries studied, the number of *BECONS* depends on the number of years, and there is one single *POCON* value (Garcia-Castro and Ariño 2016).

To interpret the results of a panel data QCA model, a researcher needs to first determine the value of POCON (Garcia-Castro and Ariño 2016). If country and time do not play a key role in determining the set-subset relation, then the value of POCON is one (Garcia-Castro and Ariño 2016). If the value of POCON is not equal to one, this means that there are inconsistencies (Garcia-Castro and Ariño 2016). These inconsistencies might be due to cases (BECONS) or years (WICONS) (Garcia-Castro and Ariño 2016). Hence, a researcher needs to study all the values in the model to understand whether the model includes inconsistent BECONS or WICONS (Garcia-Castro and Ariño 2016).

After looking through the consistency scores, we then need to analyze whether these inconsistencies are empirically relevant. We can analyze this relevance with the help of the coverage scores (Garcia-Castro and Ariño 2016). There are three main types of coverage scores: POCOV, BECOV, and

WICOV (Garcia-Castro and Ariño 2016). The **POCOV** indicates the overall/pooled coverage score, without taking the time or the country effects into consideration (Garcia-Castro and Ariño 2016). The **BECOV** (between coverage) is a measure of the cross-sectional coverage, for each year t in the panel (Garcia-Castro and Ariño 2016). The **WICOV** (within coverage) measures the set-subset relation across time, for each country in the panel (Garcia-Castro and Ariño 2016).

A researcher can also study the **BECONS distance** and **WICONS distance**. These distances range from zero to one (Garcia-Castro and Ariño 2016; Dusa 2019, Oana, Schneider, and Thomann 2021, Bhattacharya 2023a). There are four different options when analyzing the BECONS and WICONS distances (Garcia-Castro and Ariño 2016), as seen in Table 5.2.

TABLE 5.1   Types of Panel Data Consistency and Coverage Scores

MEASURES	*DEFINITION*	*NUMBER OF MEASURES*
Within consistency (WICONS)	Longitudinal consistency of the set-subset relation for each case in the panel over time, by holding the time as constant	Depends on the number of cases/countries studied
Between consistency (BECONS)	Cross-sectional consistency of the set-subset relation for each year in the panel, by holding the case as constant	Depends on the number of years studied
Pooled consistency (POCONS)	Overall consistency when the country/time is not taken into account	One
Within coverage (WICOV)	Measures the set-subset across time for each country in the panel	Depends on the number of cases/countries studied
Between Coverage (BECOV)	Measures the cross-sectional coverage for each year t in the panel	Depends on the number of years studied
Pooled Coverage (POCOV)	Overall coverage score, without taking the country or year into consideration	One

TABLE 5.2   Options While Analyzing the BECONS and WICONS Distance

- BECONS distance = WICONS distance = 0 (no evidence of case or year effects)
- BECONS distance = WICONS distance ≠ 0 (some evidence of case or year effects)
- BECONS distance > WICONS distance (year effects > cross-sectional effects)
- WICONS distance > BECONS distance (cross-sectional effects > year effects)

DATA CALIBRATION

According to this Cluster QCA approach, to calibrate data, we need to structure the dataset in long format, add two additional columns, one identifying the cases (unit_id), and one identifying the years in which each case is being studied (cluster_id) (Dusa 2022, Oana, Schneider, and Thomann 2021, Bhattacharya 2023a). As an example, if we are analyzing two or more countries over a few years, then our observations are country_year, with each country becoming a case (unit_id) and each year becoming a cluster (cluster_id). Researchers can also apply this model to analyze clusters such as geographic regions (Oana, Schneider, and Thomann 2021, 159–163).

I have not included my calibration strategy in this chapter, as I have already tabulated my calibration strategy in Table 4.7 (Chapter 4). Next, researchers need to run the necessary packages, shown below, and add row names using the pnames() function. I have added the code for pnames() as a part of my appendix. To download packages, apply the install.packages() function.

TABLE 5.3   Codes for Running Packages

```
library("readxl")
library("dplyr")
library("ggplot2")
library("QCA")
library("SetMethods")
library("readr")
library("tinytex")
library("knitr")
```

TABLE 5.4   Codes for Downloading the Dataset

```
PANEL<-read.csv("PANEL.csv")
str(PANEL)
head(PANEL)
summary(PANEL)
```

TESTING FOR NECESSARY CONDITIONS

In *Cluster QCA*, we follow the same assumptions while testing for necessary conditions as discussed in Chapter 2, but we apply the **cluster()** function to understand how the impact of the necessary condition varies over time and across the cases studied.

As mentioned in Chapter 3, in my research I have tested how "the presence of microfinance institutions" (MFISR), as a necessary condition, impacts the outcome "macro-economic participation of women" (HNWE), for the countries, Bosnia-Herzegovina, Croatia, and Montenegro, between the years 1999 and 2015. To test this, I have applied the cluster () function, seen in Tables 5.5 and 5.6. We can apply the **pof()** and **QCAfit()** functions to analyze the relevance of a necessary condition, but it shows the values of POCONS, and not WICONS and BECONS.

TABLE 5.5   Codes for Testing Presence of Microfinance Institutions (MFISR), as a Necessary Condition

```
cluster(data = PANEL, results = "MFISR", outcome =
"HNWEB",unit_id = "COUNTRY", cluster_id = "YEAR",
necessity=TRUE, wicons=TRUE)
QCAfit(PANEL$MFISR, PANEL$HNWEB, cond.lab="High Presence
of MFIs", necessity = TRUE, consH = TRUE)
```

TABLE 5.6   Codes for Testing Absence of Microfinance Institutions (MFISR), as a Necessary Condition

```
cluster(data = PANEL, results = "~MFISR", outcome =
"HNWEB",unit_id = "COUNTRY", cluster_id = "YEAR",
necessity=TRUE, wicons=TRUE)
QCAfit(1-PANEL$MFISR, PANEL$HNWEB, cond.lab="Low Presence
ofMFIs", necessity = TRUE, consH = TRUE)
```

Researchers are encouraged to test all their conditions for necessity. I have added the codes for my other variables in this chapter's appendix, at the end, and on Harvard dataverse.

INTERPRETING RESULTS FROM THE TESTING OF NECESSARY CONDITIONS

I have tabulated the results from the testing of "Presence of Microfinance Institutions, MFISR", as a necessary condition, in Table 5.7.

We need to first interpret the WICONS AND BECONS distances. My data analysis results show that the WICONS Distance > BECONS Distance (0.284 > 0.011). This means that the impact of microfinance institutions, on the national economic participation of women, varied among the countries studied more, compared to the years studied. This can also be seen through the difference between the Pooled Consistency *(POCONS)* and Between Consistency *(BECONS)* scores. There is a very small difference between Pooled Consistency (0.946) and Between Consistency scores (Between 1999-Between 2007 > 0.9), meaning that the impact of microfinance institutions was quite consistent across the years studied.

But the within consistency *(WICONS)* score for Montenegro, (within MNE: 0.226), is lower compared to the other scores. This shows that microfinance institutions did not have a significant impact in increasing national economic participation, in the case of Montenegro. In such cases, it is better for researchers to study Montenegro separately. This can be interpreted through the within to pooled consistency (0.284), which is greater than the between to pooled consistency (0.011), meaning that the effect of microfinance institutions on the national economic participation of women, does depend on the country/ cases more than the years studied (Table 5.7).

We then need to look through the between consistency and pooled consistency scores. Overall, since both the Between Consistency and the Pooled Consistency scores are greater than 0.9, the presence of microfinance institutions (MFISR) was a necessary condition in increasing the national economic participation of women (HNWEB).

So, the next step is to analyze whether the absence of Microfinance Institutions (~MFISR) still leads to an increase in economic participation of women. If the absence of Microfinance Institutions (~MFISR), still leads to an increase in economic participation of women (HNWEB), then the presence of microfinance institutions (MFISR), is **not a necessary condition** for HNWEB.

As seen in Tables 5.8 and 5.10, the absence of Microfinance Institutions (~MFISR) still leads to an increase in economic participation of women (HNWEB), in the context of the countries studied. So, microfinance was not

TABLE 5.7   Presence of MFIs (MFISR) and Economic Participation
of Women (HNWEB)

Consistencies:		
Pooled:	0.946	
Between 1999 (3)	:	0.945
Between 2000 (3)	:	0.935
Between 2001 (3)	:	0.934
Between 2002 (3)	:	0.931
Between 2003 (3)	:	0.949
Between 2004 (3)	:	0.928
Between 2005 (3)	:	0.908
Between 2006 (3)	:	1
Between 2007 (3)	:	1
Within BiH (9)	:	1
Within HRV (9)	:	1
Within MNE (9)	:	0.226
Distances:		
Between to Pooled:		0.011
Within to Pooled:		0.284
Coverages:		
Pooled:	0.307	
Between 1999 (3)	:	0.43
Between 2000 (3)	:	0.364
Between 2001 (3)	:	0.357
Between 2002 (3)	:	0.342
Between 2003 (3)	:	0.467
Between 2004 (3)	:	0.326
Between 2005 (3)	:	0.248
Between 2006 (3)	:	0.167
Between 2007 (3)	:	0.189
Within BiH (9)	:	0.225
Within HRV (9)	:	0.445
Within MNE (9)	:	0.051

TABLE 5.8 Absence of Microfinance Institutions (~MFISR) and Economic Participation of Women (HNWEB)

Consistencies:		
Pooled:	0.054	
Between 1999 (3)	:	0.055
Between 2000 (3)	:	0.065
Between 2001 (3)	:	0.066
Between 2002 (3)	:	0.069
Between 2003 (3)	:	0.051
Between 2004 (3)	:	0.072
Between 2005 (3)	:	0.092
Between 2006 (3)	:	0
Between 2007 (3)	:	0
Within BiH (9)	:	0
Within HRV (9)	:	0
Within MNE (9)	:	0.774
Distances:		
Between to Pooled:		0.191
Within to Pooled:		0.816
Coverages:		
Pooled:	0.05	
Between 1999 (3)	:	0.05
Between 2000 (3)	:	0.05
Between 2001 (3)	:	0.05
Between 2002 (3)	:	0.051
Between 2003 (3)	:	0.051
Between 2004 (3)	:	0.05
Between 2005 (3)	:	0.05
Between 2006 (3)	:	1
Between 2007 (3)	:	1
Within BiH (9)	:	1
Within HRV (9)	:	1
Within MNE (9)	:	0.05

TABLE 5.9 Codes for Using the QCAfit() Function

QCAfit(PANEL$MFISR, PANEL$HNWEB, cond.lab="High Presence of MFIs", necessity = TRUE, consH = TRUE)
QCAfit(1-PANEL$MFISR, PANEL$HNWEB, cond.lab="Low Presence ofMFIs", necessity = TRUE, consH = TRUE)

TABLE 5.10 Analyzing the Relevance of Necessary Conditions: MFISR and ~MFISR

	OUTCOME: HNWEB		
CONDITION	CONSISTENCY	COVERAGE	RELEVANCE OF NECESSITY (RON)
MFISR	0.946	0.307	0.336
~MFISR	0.0544	0.0504	0.7505

a necessary condition that increased the national economic participation of women, in these country_years.

We can also check whether Microfinance Institutions (MFISR) is a trivial or relevant necessary condition, Table 5.9, by looking at the coverage scores and using the QCAfit() function (Table 5.10).

The pooled coverage, between coverage, and within coverage scores show moderate coverage, but the coverage score of Montenegro is very low, again showing that Montenegro needs to be studied separately. The results from the QCAfit() function show that the relevance of necessary condition scores (RoN) is greater than that of coverage scores (CoV), which means the presence of Microfinance Institutions should be a relevant necessary condition.

But the results from the XYplot() function, Figure 5.1, show that there is skewness in terms of data calibration, which means the presence of Microfinance Institutions (MFISR), is a **trivial necessary condition** for HNWEB, if at all. As mentioned in Chapter 4, the calibration of Microfinance Institutions was highly skewed, and hence researchers are encouraged to recalibrate their sets if there is high skewness.

To summarize, the presence of Microfinance Institutions (MFISR), as a necessary condition, did not increase the national economic participation of women (HNWEB), in the context of the countries and years, that I have studied, but it might have combined with other factors to increase the national economic participation of women (HNWEB). To test this, we need to analyze the data analysis model for sufficiency.

Finally, a necessary condition can be a single condition, but it can also be SUIN conditions. To identify such SUIN conditions, we can also apply the superSubset() function from the QCA package (Dusa 2022; Oana, Schneider, and Thomann 2021, 80–84). But these SUIN conditions need to be theoretically justified.

TESTING FOR SUFFICIENT CONDITIONS

As discussed in Chapter 2, to test for sufficient conditions in QCA, we need to first form a truth table; second, logically minimize the truth table to remove redundant conjuncts and logically redundant prime implicants; third, create the standard analysis solutions through the conservative, parsimonious, and intermediate solutions; fourth, identify untenable assumptions (including contradictory simplifying assumptions, assumptions contradicting claims of necessity, and impossible remainders), and remove them to create the enhanced standard analysis solutions, enhanced conservative, enhanced parsimonious, and enhanced intermediate solutions (Dusa 2022, 141–214; Oana, Schneider, and Thomann 2021, 86–140). We also need to specify the directional expectations for the intermediate and enhanced intermediate solutions (Table 5.11).

In *Cluster QCA*, there is an additional step while testing for sufficient conditions. We need to apply the enhanced intermediate/parsimonious solution, as the "result" option in the cluster() function. I demonstrate each of these steps below.

Step 1: Create truth tables for the presence and absence of outcome

To create truth tables, we need to apply the truthTable() function as seen in Table 5.11.

The truth table results (Table 5.12) show that there are no logically contradictory outcomes with outcome, as "?" *or deviant cases (DCC), that lead to the presence of the outcome.* Comparing Table 5.12 and Table 5.13, shows that the combination of conditions that leads to the presence of outcome (HNWEB) and the combination of conditions that leads to the absence of outcome (~HNWEB) are different, so there are no issues related to simultaneous subset relations.

TABLE 5.11 Codes for Creating the Truth Tables

```
ttEMP<-truthTable(data=PANEL, outcome="HNWEB",
conditions= c("MFISR", "GDPR", "DEMC", "DPCL", "LTRCYR",
"LIFEXFR", "FCORR", "NEFFECTR", "FRPR", "HRTR"), incl.cut
= 0.85, sort.by = "incl,n", pri.cut=0.51, dcc= TRUE,
decreasing=FALSE,complete=TRUE, show.cases = TRUE)
ttEMP

ttemp<-truthTable(data=PANEL, outcome="~HNWEB",
conditions= c("MFISR", "GDPR", "DEMC", "DPCL", "LTRCYR",
"LIFEXFR", "FCORR", "NEFFECTR", "FRPR", "HRTR"), incl.cut
= 0.85, sort.by = "incl,n", pri.cut=0.51, dcc= TRUE,
decreasing=FALSE,complete=TRUE, show.cases = TRUE)
ttemp
```

TABLE 5.12 Truth Table for Presence of Outcome

OUT: output value
  n: number of cases in configuration
incl: sufficiency inclusion score
PRI: proportional reduction in inconsistency
DCC: deviant cases consistency

	MFISR	GDPR	DEMC	DPCL	LTRCYR	LIFEXFR	FCORR	NEFFECTR	FRPR	HRTR	OUT	n	incl	PRI
517	1	0	0	0	0	0	0	1	0	0	1	1	1.000	1.000
607	1	0	0	1	0	1	1	1	1	0	0	1	0.973	0.469
608	1	0	0	1	0	1	1	1	1	1	0	1	0.963	0.391
591	1	0	0	1	0	0	1	1	1	0	0	1	0.963	0.000
864	1	1	0	1	0	1	1	1	1	1	0	5	0.779	0.222
513	1	0	0	0	0	0	0	0	0	0	0	1	0.754	0.000
545	1	0	0	0	1	0	0	0	0	0	0	2	0.706	0.000
562	1	0	0	0	1	1	0	0	0	1	0	2	0.609	0.000
641	1	0	1	0	0	0	0	0	0	0	0	1	0.587	0.000
546	1	0	0	0	1	0	0	0	0	1	0	4	0.560	0.000
642	1	0	1	0	0	0	0	0	0	1	0	1	0.550	0.000
9	0	0	0	0	0	0	1	0	0	0	0	1	0.328	0.000
197	0	0	1	1	0	0	0	1	0	0	0	1	0.275	0.000
193	0	0	1	1	0	0	0	0	0	0	0	3	0.142	0.000
129	0	0	1	0	0	0	0	0	0	0	0	2	0.142	0.000

	DCC
517	
607	HRV01
608	HRV02
591	HRV00
864	HRV04,HRV05,HRV06,HRV07
513	BIH99
545	BIH00,BIH01
562	BIH06,BIH07
641	MNE06
546	BIH02,BIH03,BIH04,BIH05
642	MNE07
9	MNE99
197	MNE05
193	MNE02,MNE03,MNE04
129	MNE00,MNE01

TABLE 5.13 Truth Table for Absence of Outcome

OUT: output value
  n: number of cases in configuration
  incl: sufficiency inclusion score
  PRI: proportional reduction in inconsistency
DCC: deviant cases consistency

	MFISR	GDPR	DEMC	DPCL	LTRCYR	LIFEXFR	FCORR	NEFFECTR	FRPR	HRTR	OUT	n	incl	PRI	DCC
546	1	0	0	0	1	0	0	0	0	1	1	4	1.000	1.000	
193	0	0	1	1	0	0	0	0	0	0	1	3	1.000	1.000	
129	0	0	1	0	0	0	0	0	0	0	1	2	1.000	1.000	
545	1	0	0	0	1	0	0	0	0	0	1	2	1.000	1.000	
562	1	0	0	0	1	1	0	0	0	1	1	2	1.000	1.000	
9	0	0	0	0	0	0	1	0	0	0	1	1	1.000	1.000	
197	0	0	1	1	0	0	0	1	0	0	1	1	1.000	1.000	
513	1	0	0	0	0	0	0	0	0	0	1	1	1.000	1.000	
641	1	0	1	0	0	0	0	0	0	0	1	1	1.000	1.000	
642	1	0	1	0	0	0	0	0	0	1	1	1	1.000	1.000	
591	1	0	0	1	0	0	1	1	1	0	0	1	0.980	0.467	
608	1	0	0	1	0	1	1	1	1	1	1	1	0.977	0.609	
607	1	0	0	1	0	1	1	1	1	0	0	1	0.971	0.411	
864	1	1	0	1	0	1	1	1	1	1	1	5	0.937	0.778	HRV03
517	1	0	0	0	0	0	0	1	0	0	0	1	0.880	0.000	HRV99

TABLE 5.14 Codes for Logical/Boolean Minimization

```
sol_EMPB<-minimize(input=ttEMP, details=TRUE, row.dom=TRUE)
sol_EMPB
sol_empb<-minimize(input=ttemp, details=TRUE, row.dom=TRUE)
sol_empb
```

Step 2: Minimize the truth tables to remove redundant conjuncts and logically redundant prime implicants

The second step is to remove the redundant conjuncts and logically redundant prime implicants, from the truth table results.

To do this, we need to apply the minimize() function, to the truth table results, with the row.dom() option as True, shown in Table 5.14. I have tabulated the results, in Table 5.15.

Step 3: Create conservative, parsimonious, and intermediate solutions (standard analysis)

The third step is to create conservative, parsimonious, and intermediate solutions for the presence of the outcome (ttEMP), by applying the minimize()

function, to the truth table analysis results. This is known as Standard Analysis. These solutions show the causal pathways through which the presence or absence of conditions impacts the outcome in the cases studied.

I have tabulated the codes for the standard analysis in Table 5.16, and the results from the intermediate solution in Table 5.17. To create this intermediate solution, we need to first specify the directional expectations, between conditions and outcome, tabulated in Table 5.18.

TABLE 5.15 Results from the Boolean Minimization of the Truth Table

M1: MFISR*~GDPR*~DEMC*~DPCL*~LTRCYR*~LIFEXFR*~FCORR*NEFFE
CTR*~FRPR*~HRTR -> HNWEB

	inclS	PRI	covS	covU	cases
1 MFISR*~GDPR*~DEMC*~DPCL*~ LTRCYR*~LIFEXFR*~FCORR* NEFFECTR*~FRPR*~HRTR	1.000	1.000	0.210	-	HRV99
M1	1.000	1.000	0.210		

TABLE 5.16 Standard Analysis for Presence of the Outcome

```
sol_CEMP <- minimize(ttEMP, details=TRUE, row.dom=TRUE)
sol_CEMP
sol_PEMP <-minimize(ttEMP, include="?", details=TRUE,
 row.dom=TRUE)
sol_PEMP
sol_IEMP <- minimize(ttEMP, include = "?", dir.exp = "1,
 1, 1, 1, 1, 1, 1, 1, 1, 1",details=TRUE, row.dom=TRUE)
sol_IEMP
```

TABLE 5.17 Intermediate Solution

From C1P1:

M1:   MFISR*~DPCL*NEFFECTR -> HNWEB

	inclS	PRI	covS	covU	cases
1 MFISR*~DPCL*NEFFECTR	0.752	0.204	0.398	-	HRV99
M1	0.752	0.204	0.398		

From C1P2:

M1:   MFISR*~FCORR*NEFFECTR -> HNWEB

		inclS	PRI	covS	covU	cases
1	MFISR*~FCORR*NEFFECTR	0.793	0.178	0.599	-	HRV99
	M1	0.793	0.178	0.599		

From C1P3:

M1:   MFISR*NEFFECTR*~FRPR -> HNWEB

		inclS	PRI	covS	covU	cases
1	MFISR*NEFFECTR*~FRPR	0.710	0.204	0.323	-	HRV99
	M1	0.710	0.204	0.323		

From C1P4:

M1:   MFISR*~DEMC*~FCORR*NEFFECTR -> HNWEB

		inclS	PRI	covS	covU	cases
1	MFISR*~DEMC*~FCORR*NEFFECTR	0.960	0.510	0.575	-	HRV99
	M1	0.960	0.510	0.575		

From C1P5:

M1:   MFISR*~DEMC*NEFFECTR*~FRPR -> HNWEB

		inclS	PRI	covS	covU	cases
1	MFISR*~DEMC*NEFFECTR*~FRPR	1.000	1.000	0.299	-	HRV99
	M1	1.000	1.000	0.299		

TABLE 5.18   Directional Expectations Between Conditions and Outcome

CONDITIONS	INCREASES ECO. PARTICIPATION	DECREASES ECO. PARTICIPATION
MFISR	Present (1)	Absent (0)
HRTR	Present (1)	Absent (0)
GDPR	Present (1)	Absent (0)
DEMC	Present (1)	Absent (0)
DPCL	Present (1)	Absent (0)
LTRCYR	Present (1)	Absent (0)
LIFEXFR	Present (1)	Absent (0)
FCORR	Present (1)	Absent (0)
NEFFECTR	Present (1)	Absent (0)
FRPR	Present (1)	Absent (0)

Step 4: Create enhanced conservative, parsimonious, and intermediate solutions

The fourth step is to identify and remove errors, related to untenable assumptions (Oana, Schneider, and Thomann 2021). There are three main types of untenable assumptions: Contradictory Simplifying Assumptions, Assumptions Contradicting Claims of Necessity, and Assumptions on Impossible Remainders. Once we have removed these untenable assumptions, the final solution is known as Enhanced Standard Analysis (Enhanced Conservative, Enhanced Parsimonious, and Enhanced Intermediate Solution).

To identify these errors, we can use the findRows() function from the QCA package, or the esa() function from the SetMethods package. I have applied the findRows() function, as seen in Table 5.19. My results show type two errors or errors related to contradictory simplifying assumptions (CSAs). We can also apply the LR.intersect() function to identify these CSAs.

The next step is to remove these type two errors, and create enhanced solutions, by applying the exclude() option, Table 5.20. To create the enhanced intermediate solutions, we will need to apply the directional expectations, tabulated in Table 5.18.

TABLE 5.19   Codes for Identifying Errors

```
TYPETWO<-findRows(obj=ttEMP, type=2)
TYPETWO
findRows(obj=ttEMP, type=3)
```

TABLE 5.20   Codes to Create the Enhanced Solutions

```
sol_CEMPN <- minimize(ttEMP, details=TRUE, exclude=
c(TYPETWO), row.dom=TRUE)
sol_CEMPN
sol_PEMPN <- minimize(ttEMP, include="?", details=TRUE,
exclude=c(TYPETWO), row.dom=TRUE)
sol_PEMPN
sol_IEMPN <- minimize(ttEMP, include="?", details=TRUE,
dir.exp = "1, 1, 1, 1, 1, 1, 1, 1, 1, 1",
exclude=c(TYPETWO), row.dom=TRUE)
sol_IEMPN
```

TABLE 5.21   Enhanced Intermediate Solution

```
From C1P1:


M1:   MFISR*~DEMC*~DPCL*NEFFECTR*~HRTR -> HNWEB

                                   inclS    PRI    covS    covU    cases
--------------------------------------------------------------------------------
1    MFISR*~DEMC*~DPCL*NEFFECTR*~HRTR   1.000   1.000   0.371    -     HRV99
--------------------------------------------------------------------------------
                              M1   1.000   1.000   0.371



From C1P2:


M1:   MFISR*~GDPR*~DEMC*~LIFEXFR*NEFFECTR*~FRPR*~HRTR -> HNWEB

                                          inclS PRI covS covU cases
--------------------------------------------------------------------------------
1 MFISR*~GDPR*~DEMC*~LIFEXFR*NEFFECTR*~FRPR*~HRTR 1.000 1.000 0.291   -   HRV99
--------------------------------------------------------------------------------
                              M1 1.000 1.000 0.291
```

INTERPRETING THE ENHANCED INTERMEDIATE SOLUTION

The enhanced intermediate solution (sol_IEMPN, Table 5.22) shows that there are two different pathways (conjunctions) through which the presence of Microfinance Institutions (MFISR) increased the economic participation of women at the level of the national economy (HNWEB), Model 1 (C1P1) and Model 2 (C1P2). I have tabulated the enhanced intermediate solution in Table 5.22. Both these pathways were seen in Croatia during the year 1999 (cases covered: HRV99), a country which had *strong government regulations regarding the type of businesses to be set up, and where to establish microfinance programs.*

Model 1 shows that it was high government effectiveness, in terms of providing basic facilities like healthcare, education, and access to unemployment benefits, along with the presence of microfinance institutions, increased the

TABLE 5.22 Sufficient Conditions That Can Increase the Economic Participation of Women

PATH	SOLUTIONS	CONSISTENCY	PRI COVERAGE	SOL. COVERAGE	UNIQUE COVERED	CASES
Model 1:						
1	MFISR*~ DEMC*~ DPCL* NEFFECTR* ~HRTR	1.000	1.000	0.371	–	HRV99
	Overall Solution consistency: 1.000 *Overall Solution PRI: 1.000* *Overall Solution coverage: 0.371*					
Model 2:						
1	MFISR*~ GDPR* ~DEMC* ~LIFEXFR* NEFFECTR* ~FRPR*~HRTR	1.000	1.000	0.291	–	HRV99
	Overall Solution consistency: 1.000 *Overall Solution PRI: 1.000* *Overall Solution coverage: 0.291*					

economic participation of women in the case of Croatia for the year 1999, despite the absence of conditions such as high democracy (~DEMC), high political and civil liberties (~DPCL), and high number of international human rights treaties (~HRTR).

MFISR*~DEMC*~DPCL*NEFFECTR*~HRTR -> HNWEB

Model 2 shows that it was the presence of microfinance institutions (MFISR) and presence of high government effectiveness (NEFFECTR) that combined to increase the national economic participation of women, despite the absence of conditions like high GDP (~GDPR), high democracy (~DEMC), high female life expectancy (~LIFEXFR), high female representation in parliament (~FRPR), and high number of international human rights treaties (~HRTR).

MFISR*~GDPR*~DEMC*~LIFEXFR*NEFFECTR*~FRPR*~HRTR -> HNWEB

The next step in a QCA research process is to interpret why Croatia_1999 turned out to be the most typical case and how does high government effectiveness, influence the impact of microfinance on increasing the national economic participation of women (Bhattacharya 2023b, 272). For example, during the year 1999, the World Bank had just started to invest economically in rebuilding Croatia after a reduction in inflation; President Tudjman died in 1999, leading to a change in pro-market political and economic policies; there was a neighboring war in Kosovo, which might have led to a loss in employment opportunities, encouraging women in Croatia, to apply for microfinance loans, and create new employment opportunities for themselves. All of these factors could have encouraged women in Croatia to apply to Microfinance Institutions (MFISR) to increase their economic participation.

But, this doesn't explain how high government effectiveness influenced the impact of microfinance in Croatia. To understand this, researchers are encouraged to apply single case process-tracing or comparative process-tracing, to further analyze the causal pathways, with the help of the smmr() function (Oana, Schneider, and Thomann 2021, 180–200). I have added the codes for *Set-Theoretic Multi-method Research* (smmr), in Table 5.23.

TABLE 5.23 Post-QCA Set-Theoretic Multi-Method Research

```
smmr(results=sol_IEMPN, outcome="HNWEB", match=FALSE,
  cases=1, term=1)
smmr(results=sol_IEMPN, outcome="HNWEB", match=FALSE,
  cases=1, term=2)
smmr(results=sol_IEMPN, outcome="HNWEB", match=TRUE,
  cases=2, term=1)
```

Step 5: Apply the cluster function to identify how enhanced intermediate solution pathways varied across years

The final step in *Cluster QCA* is to analyze how the pathways from the enhanced intermediate solution (sol_IEMPN) varied across the cases and years studied, by applying the cluster() function.

To do this, we need to first tabulate the prime implicant membership scores from the enhanced intermediate solutions (sol_IEMPN$i.sol$C1P1$pims), as demonstrated in Table 5.24. We then need to add the prime implicant membership scores to the calibrated dataset (PANEL), Table 5.25.

TABLE 5.24 Codes for the Prime Implicant Membership Scores

```
sola<-sol_IEMPN$i.sol$C1P1$pims
Sola
solb<-sol_IEMPN$i.sol$C1P2$pims
Solb
```

TABLE 5.25 Creating a New Dataset for the cluster() Function

```
CLUSTERPANEL<-c(PANEL,sola)
CLUSTERPANEL<-as.data.frame(CLUSTERPANEL)
CLUSTERPANELB<-c(PANEL, solb)
CLUSTERPANELB<-as.data.frame(CLUSTERPANELB)
PANELA <-cluster(data = CLUSTERPANEL, results="MFISR*~DEM
C*~DPCL*NEFFECTR*~HRTR", outcome = "HNWEB",unit_id =
"COUNTRY", cluster_id = "YEAR", necessity=FALSE,
wicons=TRUE)
PANELA
PANELB <-cluster(data = CLUSTERPANELB, results= "MFISR*~G
DPR*~DEMC*~LIFEXFR*NEFFECTR*~FRPR*~HRTR", outcome =
"HNWEB",unit_id = "COUNTRY", cluster_id = "YEAR",
necessity=FALSE, wicons=TRUE)
PANELB
```

INTERPRETING DATA ANALYSIS RESULTS FROM STEP 5

As seen in Tables 5.26 and 5.27, the distances score (for between to pooled and within to pooled) is zero, which indicates that the impact of all the conditions was the same throughout the cases that I have studied. This can also be

TABLE 5.26 Results from the Cluster Function on C1P1, PANELA

Consistencies:		
Pooled:	1	
Between 1999 (3)	:	1
Between 2000 (3)	:	1
Between 2001 (3)	:	1
Between 2002 (3)	:	1
Between 2003 (3)	:	1
Between 2004 (3)	:	1
Between 2005 (3)	:	1
Between 2006 (3)	:	1
Between 2007 (3)	:	1
Within BiH (9)	:	1
Within HRV (9)	:	1
Within MNE (9)	:	1
Distances:		
Between to Pooled:		0
Within to Pooled:		0
Coverages:		
Pooled:	0.371	
Between 1999 (3)	:	0.668
Between 2000 (3)	:	0.614
Between 2001 (3)	:	0.204
Between 2002 (3)	:	0.187
Between 2003 (3)	:	0.186
Between 2004 (3)	:	0.333
Between 2005 (3)	:	0.375
Between 2006 (3)	:	0.516
Between 2007 (3)	:	0.256
Within BiH (9)	:	0.472
Within HRV (9)	:	0.361
Within MNE (9)	:	0.007

TABLE 5.27 Results from the Cluster Function on C1P2, PANELB

Consistencies:

Pooled:	1	
Between 1999 (3)	:	1
Between 2000 (3)	:	1
Between 2001 (3)	:	1
Between 2002 (3)	:	1
Between 2003 (3)	:	1
Between 2004 (3)	:	1
Between 2005 (3)	:	1
Between 2006 (3)	:	1
Between 2007 (3)	:	1
Within BiH (9)	:	1
Within HRV (9)	:	1
Within MNE (9)	:	1

Distances:

Between to Pooled:	0
Within to Pooled:	0

Coverages:

Pooled:	0.291	
Between 1999 (3)	:	0.668
Between 2000 (3)	:	0.195
Between 2001 (3)	:	0.199
Between 2002 (3)	:	0.182
Between 2003 (3)	:	0.182
Between 2004 (3)	:	0.27
Between 2005 (3)	:	0.295
Between 2006 (3)	:	0.413
Between 2007 (3)	:	0.189
Within BiH (9)	:	0.472
Within HRV (9)	:	0.232
Within MNE (9)	:	0.007

concluded from the fact that the Pooled Consistency and Between Consistency scores are all 1.

The pooled coverage, between coverage, and within coverage scores show moderate coverage in all cases, except Montenegro, where the extremely low coverage score indicates that my data analysis model is not a good fit, similar to the results from the testing of MFISR, as a necessary condition.

To visualize the cluster analysis results from the testing of sufficient conditions, we can apply the cluster.plot() function, along with the XYplot() function. For example, by applying the cluster.plot() function to the enhanced intermediate solution, we can visualize if the sufficient pathway from C1P1 varies across the clusters BiH, MNE, and HRV. I have added the code in Table 5.28, and visualized the cluster analysis result, in Figure 5.3. Since my within consistency scores were all 1, Figure 5.3 shows that the pathway did not vary much amongst the clusters.

TABLE 5.28   Codes for the Cluster.plot() Function

```
CD <-cluster(data=PANEL, results=sol_IEMPN, outcome=
 "HNWEB", unit_id="COUNTRY", cluster_id="YEAR", necessity
 = FALSE, wicons=TRUE)
cluster.plot(cluster.res= CD, labs=TRUE, size=7, angle=15,
 wicons=TRUE, wiconslabs = TRUE)
```

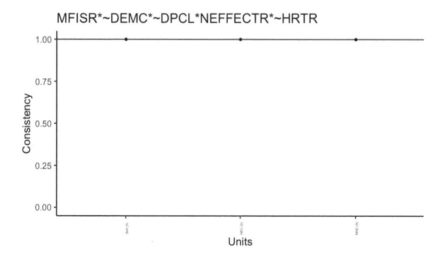

VISUALIZING RESULTS FROM TESTING OF NECESSARY AND SUFFICIENT CONDITIONS

To visualize the pathways from the testing of necessary and sufficient conditions, we can apply the XYplot(), xy.plot(), or pimplot() functions, either on the entire solution/conjunction or on each individual pathway/conjunct.

XYplot() can also help researchers to identify typical cases, deviant cases consistency in kind, deviant cases consistency in degree, irrelevant cases, and individually irrelevant cases (Rohlfing and Schneider 2013, 2018; Schneider and Rohlfing 2016; Williams and Gemperle 2017). I have tabulated the codes for testing MFISR as a necessary condition in Table 5.29, and for visualizing C1P1 and C1P2 as sufficient pathways in Table 5.30. I have also added the figures in Figure 5.2 (testing MFISR as a necessary condition) and Figure 5.3 (visualizing C1P1 and C1P2 as sufficient pathways).

TABLE 5.29   Codes for the XY Plots, Testing MFISR as a Necessary Condition

```
CLUSTERPANEL$Observations<-NULL

CLUSTERPANEL$COUNTRY<-NULL

CLUSTERPANEL$YEAR<-NULL

CLUSTERPANEL$X<-NULL

XYplot(MFISR, HNWEB, data=CLUSTERPANEL, relation =
 "necessity", enhance = TRUE, jitter=TRUE, clabels =
 seq(nrow(CLUSTERPANEL)))
```

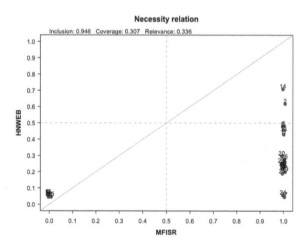

TABLE 5.30 Codes for the XY Plots, Testing C1P1 and C1P2 as Sufficient Conditions

```
XYplot(MFISR*~DEMC*~DPCL*NEFFECTR*~HRTR, HNWEB,
 data=CLUSTERPANEL, relation = "sufficiency", enhance =
 TRUE, jitter=TRUE, clabels = seq(nrow(CLUSTERPANEL)))
XYplot(MFISR*~GDPR*~DEMC*~LIFEXFR*NEFFECTR*FRPR*~HRTR,
 HNWEB,data=CLUSTERPANEL, relation = "sufficiency",
 enhance = TRUE, jitter=TRUE, clabels =
 seq(nrow(CLUSTERPANEL)))
```

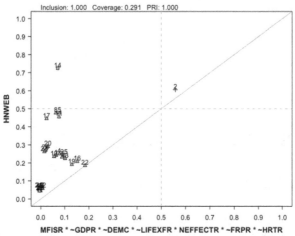

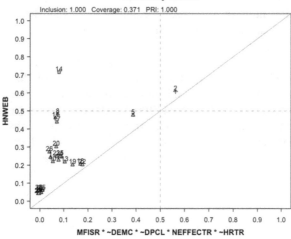

POST-QCA ANALYSIS

Along with the steps discussed in chapter 2, researchers are also encouraged to test for robustness.

I encourage readers to follow the suggestions offered by Oana, Schneider, and Thomann (2021, 146–158) in their detailed discussion of "robustness in QCA," for all the four approaches demonstrated in this book. As an example, I applied the rob.calibrange () function to test the calibration robustness of GDPR, for this chapter's data analysis model. I have listed the codes in the chapter appendix.

Researchers are also encouraged to test the applicability of their theories in the context of the cases studied. To do this, researchers can apply the theory. evaluation() or modelFit() functions. I have listed both these codes in the chapter appendix.

STRENGTHS OF CLUSTER QCA

A few advantages of this approach are (Bhattacharya 2023b, 273–274):

- It allows researchers to study how macrosocial units have developed over time and across cases, and whether that change is due to the years (BECONS) or countries studied (WICONS).
- Like cross-sectional QCA, it also allows researchers to study a smaller number of cases, as compared to panel data regression models, and to include data from various levels of analysis, such as micro, meso, and macro.
- By combining features from both qualitative and quantitative research, panel data QCA broadens the range of data analysis approaches available to mixed-methods researchers. For example, Garcia-Castro and Ariño (2016) developed *Cluster QCA* model based on panel data econometrics, but to build such a model, researchers must understand the cases studied, include variables from the case study, calibrate variables based on the concepts of relevant variation and qualitative change, and interpret the data analysis results based on the case study – all tasks usually associated with qualitative research.

WEAKNESSES

Despite these advantages, in the last few years, scholars of QCA have cited a few disadvantages (Bhattacharya 2023, 274). For example, in QCA, researchers are encouraged to study "qualitative change over time," but the panel data model suggested by Garcia-Castro and Ariño (2016) may not be able to capture this change, due to its repeated measure arrangement.

To address this critique, in the next few chapters, I will demonstrate three alternative panel data QCA models.

Appendix

Dataset: PANEL.CSV

OBSERVATIONS	PRESENCEOFMFIS	NUMBEROFMFIS	GDP	DEM	PCL	LTRCYF	LIFEXF	FCOR
Bosnia and Herzegovina_1999	1	18	4.69E+09	−66	5	98.73545	76.581	NA
Croatia_1999	1	3	2.34E+10	0	4	99.6044	76.55	NA
Montenegro_1999	0	4	1.84E+10	0	5	98.98921	76.11	NA
Bosnia and Herzegovina_2000	1	18	5.51E+09	−66	4.5	99.84363	76.881	33.50254
Croatia_2000	1	3	2.17E+10	0	2.5	99.6044	76.68	56.34518
Montenegro_2000	0	4	9.84E+08	7	4	98.98921	76.035	53.29949
Bosnia and Herzegovina_2001	1	18	5.75E+09	−66	4.5	99.84363	77.128	NA
Croatia_2001	1	3	2.32E+10	0	2	99.65321	78.17	NA
Montenegro_2001	0	4	1.16E+09	7	3	98.98921	76.022	NA
Bosnia and Herzegovina_2002	1	18	6.65E+09	−66	4	99.84363	77.333	42.92929
Croatia_2002	1	3	2.68E+10	0	2	99.65321	78.4	63.63636
Montenegro_2002	0	4	1.28E+09	7	2.5	98.98921	76.04	53.53535

Bosnia and Herzegovina_2003	1	18	8.37E+09	−66	4	99.84363	77.508	46.46465
Croatia_2003	1	3	3.47E+10	0	2	99.65321	78.23	64.14141
Montenegro_2003	0	4	1.71E+09	6	2.5	99.3606	76.068	41.91919
Bosnia and Herzegovina_2004	1	18	1.06E+10	−66	3.5	99.84363	77.659	44.87805
Croatia_2004	1	3	4.16E+10	0	2	99.65321	79.08	62.92683
Montenegro_2004	0	4	2.07E+09	6	2.5	99.3606	76.106	38.04878
Bosnia and Herzegovina_2005	1	18	1.12E+10	−66	3.5	99.84363	77.793	50.73171
Croatia_2005	1	3	4.54E+10	0	2	99.65321	78.83	58.53659
Montenegro_2005	0	4	2.26E+09	6	2.5	99.3606	76.172	45.85366
Bosnia and Herzegovina_2006	1	18	1.29E+10	−66	3	99.84363	77.92	44.39024
Croatia_2006	1	3	5.05E+10	0	2	99.65321	79.33	59.5122
Montenegro_2006	1	4	2.72E+09	9	3	99.3606	76.291	40
Bosnia and Herzegovina_2007	1	18	1.58E+10	−66	3.5	99.84363	78.047	41.26213
Croatia_2007	1	3	6.02E+10	0	2	99.65321	79.25	58.73787
Montenegro_2007	1	4	3.68E+09	9	3	99.3606	76.484	47.08738

(Continued)

OBSERVATIONS	FEFFECT	FPARL	EUM	INTHRT	OWN	FAM	EMPE	EMPR	COUNTRY	YEAR
Bosnia and Herzegovina_1999	NA	NA	0	8	49876	59102	241554	10745	BiH	1999
Croatia_1999	NA	7.9	0	8	82003	68965	574492	24269	HRV	1999
Montenegro_1999	NA	NA	0	0	5022	2040	55634	3255	MNE	1999
Bosnia and Herzegovina_2000	17.94872	NA	0	8	48935	54411	243877	10776	BiH	2000
Croatia_2000	66.66666	20.5	0	8	75374	60513	561296	22884	HRV	2000
Montenegro_2000	NA	NA	0	0	4896	2042	55602	3281	MNE	2000
Bosnia and Herzegovina_2001	NA	7.1	0	9	47803	50548	244760	10757	BiH	2001
Croatia_2001	NA	20.5	0	9	74852	55749	563222	17896	HRV	2001
Montenegro_2001	NA	NA	0	0	4820	1976	55812	3331	MNE	2001
Bosnia and Herzegovina_2002	13.77551	NA	0	11	46485	47317	244288	10685	BiH	2002
Croatia_2002	65.81633	20.5	0	11	72761	54332	556225	26295	HRV	2002
Montenegro_2002	NA	NA	0	0	4727	1919	55936	3371	MNE	2002
Bosnia and Herzegovina_2003	22.95918	16.7	1	12	45518	44287	245223	10681	BiH	2003
Croatia_2003	67.34694	NA	1	11	97368	48321	559658	20141	HRV	2003
Montenegro_2003	NA	NA	0	0	4650	1865	56292	3421	MNE	2003
Bosnia and Herzegovina_2004	29.55665	16.7	1	12	44543	41665	245829	10665	BiH	2004

Case										
Croatia_2004	68.96552	21.7	2	11	113331	39525	578198	21062	HRV	2004
Montenegro_2004	NA	NA	0	0	4600	1811	56861	3450	MNE	2004
Bosnia and Herzegovina_2005	25.98039	16.7	1	12	42294	37994	239891	10302	BiH	2005
Croatia_2005	67.15686	21.7	2	12	134432	30940	582765	21949	HRV	2005
Montenegro_2005	64.21568	NA	0	0	4569	1750	57521	3469	MNE	2005
Bosnia and Herzegovina_2006	32.19512	14.3	1	12	41522	35842	240758	10308	BiH	2006
Croatia_2006	69.7561	21.7	2	12	112857	28208	614741	24806	HRV	2006
Montenegro_2006	52.68293	8.6	0	10	5713	2010	64231	4370	MNE	2006
Bosnia and Herzegovina_2007	20.38835	11.9	1	12	49631	29957	250431	10485	BiH	2007
Croatia_2007	68.93204	20.9	2	14	96374	28923	629619	22952	HRV	2007
Montenegro_2007	47.57281	11.1	0	11	6861	2204	69702	5287	MNE	2007

Codes for testing the remaining conditions as necessary, presence of necessary
conditions

```
cluster(data = PANEL, results = "GDPR", outcome =
"HNWEB",unit _ id = "COUNTRY", cluster _ id = "YEAR",
necessity=TRUE, wicons=TRUE)

cluster(data = PANEL, results = "DEMC", outcome =
"HNWEB",unit _ id = "COUNTRY", cluster _ id = "YEAR",
necessity=TRUE, wicons=TRUE)

cluster(data = PANEL, results = "DPCL", outcome =
"HNWEB",unit _ id = "COUNTRY", cluster _ id = "YEAR",
necessity=TRUE, wicons=TRUE)

cluster(data = PANEL, results = "LTRCYR", outcome =
"HNWEB",unit _ id = "COUNTRY", cluster _ id = "YEAR",
necessity=TRUE, wicons=TRUE)

cluster(data = PANEL, results = "LIFEXFR", outcome =
"HNWEB",unit _ id = "COUNTRY", cluster _ id = "YEAR",
necessity=TRUE, wicons=TRUE)

cluster(data = PANEL, results = "FCORR", outcome =
"HNWEB",unit _ id = "COUNTRY", cluster _ id = "YEAR",
necessity=TRUE, wicons=TRUE)

cluster(data = PANEL, results = "NEFFECTR", outcome
= "HNWEB",unit _ id = "COUNTRY", cluster _ id = "YEAR",
necessity=TRUE, wicons=TRUE)

cluster(data = PANEL, results = "FRPR", outcome =
"HNWEB",unit _ id = "COUNTRY", cluster _ id = "YEAR",
necessity=TRUE, wicons=TRUE)

cluster(data = PANEL, results = "HRTR", outcome =
"HNWEB",unit _ id = "COUNTRY", cluster _ id = "YEAR",
necessity=TRUE, wicons=TRUE)
```

Codes for testing the remaining conditions as necessary, absence of necessary
conditions

```
cluster(data = PANEL, results = "~GDPR", outcome =
"HNWEB",unit _ id = "COUNTRY", cluster _ id = "YEAR",
necessity=TRUE, wicons=TRUE)

cluster(data = PANEL, results = "~DEMC", outcome =
"HNWEB",unit _ id = "COUNTRY", cluster _ id = "YEAR",
necessity=TRUE, wicons=TRUE)
```

```
cluster(data = PANEL, results = "~DPCL", outcome =
"HNWEB",unit_id = "COUNTRY", cluster_id = "YEAR",
necessity=TRUE, wicons=TRUE)
cluster(data = PANEL, results = "~LTRCYR", outcome =
"HNWEB",unit_id = "COUNTRY", cluster_id = "YEAR",
necessity=TRUE, wicons=TRUE)
cluster(data = PANEL, results = "~LIFEXFR", outcome
= "HNWEB",unit_id = "COUNTRY", cluster_id = "YEAR",
necessity=TRUE, wicons=TRUE)
cluster(data = PANEL, results = "~FCORR", outcome =
"HNWEB",unit_id = "COUNTRY", cluster_id = "YEAR",
necessity=TRUE, wicons=TRUE)
cluster(data = PANEL, results = "~NEFFECTR", outcome
= "HNWEB",unit_id = "COUNTRY", cluster_id = "YEAR",
necessity=TRUE, wicons=TRUE)
cluster(data = PANEL, results = "~FRPR", outcome =
"HNWEB",unit_id = "COUNTRY", cluster_id = "YEAR",
necessity=TRUE, wicons=TRUE)
cluster(data = PANEL, results = "~HRTR", outcome =
"HNWEB",unit_id = "COUNTRY", cluster_id = "YEAR",
necessity=TRUE, wicons=TRUE)
```

Codes for theory evaluation/theory-testing:

```
modelFit(sol_IEMPN, "MFISR*GDPR*DEMC*DPCL*LTRCYR*LIFEX
FR*FCORR*NEFFECTR*FRPR*HRTR")
t<-"MFISR*GDPR*DEMC*DPCL*LTRCYR*LIFEXFR*FCORR*NEFFECTR*
FRPR*HRTR"
THEVAL<- theory.evaluation(theory=t, empirics=sol_
IEMPN, outcome="HNWEB", sol="c1p1i1", print.data=TRUE,
print.fit=TRUE)
THEVAL
```

Codes for testing robustness in calibration range:

```
CALIB<-PANEL %>% select(X, Observations, MFISR, GDPR,
DEMC, DPCL, LTRCYR, LIFEXFR, FCORR, NEFFECTR, FRPR,
HRTR, OWNR, FAMR, EMPER, EMPRR)
```

```
RAW<-DATA %>% select(X, Observations, PresenceofMFIs,
GDP, DEM, PCL, LTRCYF, LIFEXF, FCOR, FEFFECT, FPARL,
INTHRT, OWN, FAM, EMPE, EMPR)
conds<-c("MFISR", "GDPR","DEMC", "DPCL", "LTRCYR",
"LIFEXFR", "FCORR", "NEFFECTR", "FRPR", "HRTR")
condsr<-c("PresenceofMFIs", "GDP","DEM", "PCL",
"LTRCYF", "LIFEXF", "FCOR", "FEFFECT", "FPARL",
"INTHRT")
rob.calibrange(raw.data=RAW,
calib.data=CALIB,
test.cond.raw="GDP",
test.cond.calib = "GDPR",
test.thresholds=c(980000000, 30000000000, 50000000000),
type="fuzzy",
step=50000,
max.runs=40,
outcome="OWNR",
conditions = conds,
incl.cut=0.85,
n.cut=1,
include="?")
```

Multiple Sub-QCA

6

Chapter outline:

- **Sub-QCA Model A:-**
  - Introduction
  - Assumptions
  - My data analysis model and calibration
  - Testing for necessary conditions
  - Testing for sufficient conditions
  - Interpreting the results from the enhanced intermediate solution
  - Visualizing necessary and sufficient conditions

- **Sub-QCA Model B:-**
  - Testing for necessary conditions
  - Testing for sufficient conditions
  - Visualizations
  - Strengths and weaknesses

- Appendix
- Sub-QCA Model A, year 1999 supplementary codes
- Sub-QCA Model B, years 1999 and 2007

INTRODUCTION

In this chapter, I will discuss the *Multiple Sub-QCA* approach, based on the models suggested by Verweij and Vis (2021). This approach can help us analyze how the impact of conditions has changed over time and across cases, while also maintaining "qualitative change over time." To ensure this:

- First, researchers can set up separate QCA models for different years, usually at a gap of five to ten years. I have discussed this as Sub-QCA Model A.

As an example for Chapter 5's data analysis model, I can set up separate cross-sectional QCA models for the years 1999 and 2007.

Second, researchers can increase the number of cases studied by including different time-points in a single QCA model (Verweij and Vis 2021, 100–103). I have discussed this approach as Sub-QCA Model B.

As an example, I can combine the years 1999 and 2008, in one QCA model, and get a total of six cases, BiH_1999, HRV_1999, MNE_1999, BiH_2007, HRV_2007, and MNE_2007.

ASSUMPTIONS

The assumptions for Models A and B are the same as those of cross-sectional QCA: that is, causal asymmetry, conjunctural causation, equifinality, and testing for necessary conditions before testing for sufficient conditions.

MY DATA ANALYSIS MODEL AND CALIBRATION

As mentioned in Chapter 4, since this chapter's models are different from those in Chapter 5, I have calibrated my data separately for each of the two models separately, in Tables 6.1 and 6.15.

Since my case studies had not yet started their process of European Union (EU) membership, I did not include EU membership as a condition for the models discussed in this chapter, as well.

TABLE 6.1 Model A (Year: 1999)

CONCEPT	DIMENSION/ CHARACTERISTIC	INDICATOR/ VARIABLE	ABBREVIATION	TYPE OF SET	TYPE OF CALIBRATION	SET LABEL
Microfinance Institutions	Presence of microfinance institutions	Presence of MFIs **Source:** World Bank| Databank| MIX Market	MFISR	Crisp()	(Threshold=1)	Set of countries with presence of microfinance institutions
Access to Political Freedom	Extent of democracy within country	Democracy Scale (DEM), **Source:** Polity IV Dataset	DEMC	Crisp()	Crisp-Set (threshold=0)	Set of countries with high extent of democracy
	Access to political and civil rights	Political Rights & Civil Liberty Scale (PCL) **Source:** Freedom Rating Scale, Freedom House Reports	DPCL	Crisp()	Crisp-Set (recode, 5=0, else=1)	Set of countries with high access to political and civil rights
Access to Economic Facilities	Economic growth – > increases opportunities available to individuals for production, consumption, and exchange	Gross Domestic Product **Source:** World Bank	GDPR	Fuzzy()	Direct Calibration Method (thresholds, "e=460000000, c= 2000000000, i= 23000000000")	Set of countries with high economic growth

(Continued)

TABLE 6.1 (Continued)

CONCEPT	DIMENSION/ CHARACTERISTIC	INDICATOR/ VARIABLE	ABBREVIATION	TYPE OF SET	TYPE OF CALIBRATION	SET LABEL
Access to Social Opportunities	Access to basic education and healthcare, which will help an individual live a better life	% of literate females ages 15–24 **Source:** World Bank	LTRCYR	Fuzzy()	Direct Calibration Method (thresholds= e=98.70, c=99.00, i=99.60)	Set of countries with high percentage of literate females
		Life expectancy at birth for females, measured in years **Source:** World Bank	LIFEXFR		Direct Calibration Method (thresholds= e=76.10, c=76.56, i=76.58)	Set of countries with high life expectancy for females
Access to Transparency Guarantees	Society operates on the basic presumption of trust. Need for openness, guarantees of disclosure, and low corruption	Control of Corruption Scale **Source:** World Governance Indicators| World Bank	NCORR	Fuzzy()	Direct Calibration Method (thresholds, e=34.0, c=45.0, i=71.0)	Set of countries with high control of corruption
Access to Protective Security	Effectiveness of public institutions in ensuring access to basic facilities like sanitation, health, and unemployment benefits	Government Effectiveness Scale **Source:** World Governance Indicators| World Bank	NEFFECTR	Fuzzy()	Direct Calibration Method (Thresholds, e=0.0, c=15.0, i=59.0)	Set of countries with high government effectiveness

International Human Rights Treaties (Case Study Condition)	Signed human rights treaties to eliminate all forms of racial and gender inequality	Signed 18 international human rights treaties **Source:** UN Human Rights Office of the High Commissioner	HRTR	Crisp() (threshold=8)	Set of countries with high number of international human rights treaties signed
Female Representation in Parliament (Case Study Condition)	Percentage of seats held by women in national parliaments	Proportion of seats held by women in national parliaments (%) **Source:** Millennium Development Goals, World Bank	FRPR	Crisp() (threshold=7.9)	Set of countries with high female representation in parliament
Economic Empowerment of Women (Outcome)	Economic participation of women at the level of the national economy, for the economic sectors of own-account workers (informal economy), family workers (informal economy), employees (formal economy), employers (formal economy)	Number of employed women ages 15–65 **Source:** Employment by Sex and Status in Employment (KILM 3); ILO Modeled Estimates – Annual, calculated in thousands	Macro-condition: (HNWEB) OWNA, FAMA, EMPEA, EMPRA	fuzzyand() Direct Calibration Method OWNR: (thresholds=" e=5000, c=51000, i=80000") FAMR: (thresholds= "e=2000, c=61000, i=68000") EMPRR: (thresholds= "e=55000, c=260000, i=570000") EMPRA: (thresholds= "e=3200, c=12000, i=24000")	Set of countries with high economic participation of women

TESTING FOR NECESSARY CONDITIONS

To test for necessary conditions, we can apply the pof() or the QCAfit() functions. I have applied the QCAfit() function to test for the presence and absence of necessary conditions (Table 6.2). According to the assumptions of QCA, a necessary condition should have a consistency score of at least 0.9 and a coverage score of at least 0.5, and the RoN() score should be greater than the coverage score if it is a relevant necessary condition.

My data analysis results show that "high presence of MFIs" (MFISR), "high government effectiveness" (NEFFECTR), and "high number of international human rights treaties signed" (HRTR), are necessary conditions for the presence of the outcome "high national economic participation by women" (HNWEB), Table 6.3. However, the absence of these necessary conditions does not lead to the absence of the outcome, HNWEB, Table 6.3. Therefore, these are trivial necessary conditions. We can also test for SUIN necessary conditions by applying the superSubset() function.

Before testing for necessary conditions, we will need to download the dataset "FinalData_1999.csv," install and run the required RStudio packages, and calibrate the variables. I have tabulated the codes for the testing of MFIS as a necessary condition in Table 6.2 and added the dataset, the codes for testing the other conditions, in the chapter appendix and on HarvardDataverse.

TABLE 6.2   Codes for Testing the Presence and Absence of MFIs as a Necessary Condition

```
QCAfit (DATA$MFISR, DATA$HNWEB, cond.lab="Presence of
 MFIs", necessity = TRUE, consH = TRUE)
QCAfit(1-DATA$MFISR, DATA$HNWEB, cond.lab="Absence of
 MFIs", necessity = TRUE, consH = TRUE)
```

TABLE 6.3   Data Analysis Results for the Year 1999

CONDITIONS	INCLN	COVN	RON
MFISR	0.964	0.674	0.606
~MFISR	0.0358	0.0501	0.678
GDPR	0.753	0.731	0.801
~GDPR	0.344	0.308	0.572

CONDITIONS	INCLN	COVN	RON
DEMC	0.716	0.501	0.501
~DEMC	0.284	0.397	0.768
DPCL	0.681	0.952	0.977
~DPCL	0.319	0.223	0.392
LTRCYR	0.765	0.717	0.781
~LTRCYR	0.354	0.329	0.597
LIFEXF	0.665	0.623	0.728
~LIFEXF	0.436	.405	0.625
NCORR	0.331	0.340	0.645
~NCORR	0.993	0.849	0.846
NEFFECTR	0.952	1.000	1.000
~NEFFECTR	0.354	0.297	0.532
FRPR	0.681	0.952	0.977
~FRPR	0.319	0.223	0.392
HRTR	0.964	0.674	0.606
~HRTR	0.036	0.050	0.678

TESTING FOR SUFFICIENT CONDITIONS

As described in the last chapter, to test for sufficient conditions, we need to first create truth tables for both the presence and the absence of outcome. I have applied the truthTable() function to create these truth tables, Table 6.4. We then need to logically minimize them to remove logically redundant primitive expressions and logically redundant prime implicants. The third step is to create conservative, parsimonious, and intermediate solutions, as a part of standard analysis. The final step is to create an enhanced standard analysis and remove the rows with untenable assumptions. We also need to specify the *directional expectations* for intermediate and enhanced intermediate solutions.

To avoid repetition, I have listed the codes for these steps below, but I discuss only the results from the enhanced intermediate solution, to compare it with Chapter 5.

TABLE 6.4 Directional Expectations between Conditions and Outcome

CONDITIONS	INCREASES ECO. PARTICIPATION	DECREASES ECO. PARTICIPATION
MFISR	Present (1)	Absent (0)
HRTR	Present (1)	Absent (0)
GDPR	Present (1)	Absent (0)
DEMC	Present (1)	Absent (0)
DPCL	Present (1)	Absent (0)
LTRCYR	Present (1)	Absent (0)
LIFEXFR	Present (1)	Absent (0)
NCORR	Present (1)	Absent (0)
NEFFECTR	Present (1)	Absent (0)
FRPR	Present (1)	Absent (0)

Step 1: Create truth tables

TABLE 6.5 Code for Creating the Truth Tables

```
ttEMP<-truthTable(data=PANELB, outcome="HNWEB",
conditions= c("MFISR", "GDPR", "DEMC", "DPCL",
"LTRCYR", "LIFEXFR", "NCORR", "NEFFECTR", "FRPR",
"HRTR"), incl.cut = 0.85, sort.by = "incl,n", pri.
cut=0.51, dcc= TRUE, decreasing=FALSE,complete=TRUE,
show.cases = TRUE)

ttEMP

ttemp<-truthTable(data=PANELB, outcome="~HNWEB",
conditions= c("MFISR", "GDPR", "DEMC", "DPCL", "LTRCYR",
"LIFEXFR", "NCORR", "NEFFECTR", "FRPR", "HRTR"), incl.cut
= 0.85, sort.by = "incl,n", pri.cut=0.51, dcc= TRUE,
decreasing=FALSE,complete=TRUE, show.cases = TRUE)

ttemp
```

Comparing Tables 6.6 and 6.7, it is evident that different combinations of conditions cause the presence and absence of the outcome, which means there are no issues with simultaneous subset relations. Table 6.6 also shows that there are no DCC cases associated with the presence of the outcome. So, the next step is to minimize the truth tables.

TABLE 6.6   Truth Table for Presence of the Outcome

OUT: output value
   n: number of cases in configuration
  incl: sufficiency inclusion score
  PRI: proportional reduction in inconsistency
DCC: deviant cases consistency

	MFISR	GDPR	DEMC	DPCL	LTRCYR	LIFEXFR	NCORR	NEFFECTR	FRPR	HRTR	OUT	n	incl	PRI
1000	1	1	1	1	1	0	0	1	1	1	1	1	1.000	1.000
530	1	0	0	0	0	1	0	0	0	1	0˙	1	0.616	0.000
137	0	0	1	0	0	0	1	0	0	0	0	1	0.095	0.000
	DCC													
1000														
530	BIH99													
137	MNE99													

TABLE 6.7   Truth Table for Absence of the Outcome

OUT: output value
   n: number of cases in configuration
  incl: sufficiency inclusion score
  PRI: proportional reduction in inconsistency
DCC: deviant cases consistency

	MFISR	GDPR	DEMC	DPCL	LTRCYR	LIFEXFR	NCORR	NEFFECTR	FRPR	HRTR	OUT	n	incl	PRI
137	0	0	1	0	0	0	1	0	0	0	1	1	1.000	1.000
530	1	0	0	0	0	1	0	0	0	1	1	1	0.937	0.836
1000	1	1	1	1	1	0	0	1	1	1	0	1	0.093	0.000
	DCC													
137														
530														
1000	HRV99													

TABLE 6.8   Codes for Boolean Minimization

```
sol_EMPB<-minimize(input=ttEMP, details=TRUE, row.dom=TRUE)
sol_EMPB
sol_empb<-minimize(input=ttemp, details=TRUE, row.dom=TRUE)
sol_empb
```

Step 2: Minimize the truth tables to remove redundant conjuncts and logically redundant prime implicants

To remove redundant conjuncts and logically redundant prime implicants, we need to apply the minimize() function, with row.dom() option. I have tabulated the codes in Table 6.8.

Step 3: Standard Analysis: Create the conservative, parsimonious, and intermediate solutions

To create the conservative, parsimonious, and intermediate solutions, we need to apply the minimize() function, to the presence as well as the absence of the outcome, as seen in Tables 6.9 and 6.10. We also need to apply the directional expectations for the intermediate solution.

Step 4: Create enhanced conservative, parsimonious, and intermediate solutions: Enhanced Standard Analysis

To create the enhanced conservative, parsimonious, and intermediate solutions, we need to first identify and remove errors. I applied the findRows() function to do this, Table 6.11.

TABLE 6.9   Standard Analysis, Presence of the Outcome

```
sol_CEMP <- minimize(ttEMP, details=TRUE, row.dom=TRUE)
sol_CEMP
sol_PEMP <-minimize(ttEMP, include="?", details=TRUE,
 row.dom=TRUE)
sol_PEMP
sol_IEMP <- minimize(ttEMP, include = "?", dir.exp = "1,
 1, 1, 1, 1, 1, 1, 1, 1, 1",details=TRUE, row.dom=TRUE)
sol_IEMP
```

TABLE 6.10   Standard Analysis, Absence of the Outcome

```
sol_cemp <- minimize(ttemp, details=TRUE, row.dom=TRUE)
sol_cemp
sol_pemp<-minimize(ttemp, include="?", details=TRUE, row.
 dom=TRUE)
sol_pemp
sol_iemp <- minimize(ttemp, include = "?", dir.exp = "0,
 0, 0, 0, 0, 0, 0, 0, 0, 0",details=TRUE, row.dom=TRUE)
sol_iemp
```

TABLE 6.11   Identify Errors

```
findRows(obj=ttHNWEA, type=0)
TYPETWO<-findRows(obj=ttHNWEA, type=2)
TYPETWO
findRows(obj=ttHNWEA, type=3)
```

TABLE 6.12   Codes for Creating the Enhanced Conservative, Parsimonious, and Intermediate Solutions

```
sol_CEMPN<-minimize(ttEMP, details=TRUE, exclude=c(TYPETWO))
sol_CEMPN
sol_PEMPN<-minimize(ttEMP, include="?", details=TRUE,
 exclude=c(TYPETWO))
sol_PEMPN
sol_IEMPN<-minimize(ttEMP, include = "?", dir.exp = "1, 1,
 1, 1, 1, 1, 1, 1, 1, 1", details=TRUE, exclude=c(TYPETWO))
sol_IEMPN
```

TABLE 6.13   Enhanced Intermediate Solution

```
From C1P1:
```

M1: MFISR*GDPR*DEMC*DPCL*LTRCYR*NEFFECTR*FRPR*HRTR -> HNWEB

		inclS	PRI	covS	covU	cases
1	MFISR*GDPR*DEMC*DPCL*LTRCYR*NEFF ECTR*FRPR*HRTR	1.000	1.000	0.680	–	HRV99
	M1	1.000	1.000	0.680		

Since there are no type 0, 1, or 3 errors, I have applied the "QCA" package to remove *the type two* errors and create the enhanced conservative, parsimonious, and intermediate solutions, Table 6.12. I have tabulated the results from the enhanced intermediate solution, in Table 6.13.

INTERPRETING THE RESULTS FROM THE ENHANCED INTERMEDIATE SOLUTION

The enhanced intermediate solution (Table 6.13) shows that strong political, economic, and social factors, like high economic growth (GDPR), high extent of democracy (DEMC), high political and civil liberty (DPCL), high extent of female literacy (LTRCYR), high government effectiveness (NFEFFECTR), high female representation in parliament (FRPR), and high presence of international human rights (HRTR), increased the efffectiveness of microfinance on increasing the national economic participation of women, in Croatia, for the year 1999. The next step is to analyze whether this pathway changed for another year, for example, 2007, or whether any new pathways might have increased the economic participation of women in the context of Bosnia-Herzegovina, Croatia, and Montenegro.

Comparing this enhanced intermediate solution (Table 6.13), with the one in Chapter 5 (Table 5.22), shows that the influence of these conditions, like presence of high democracy, high political and civil liberty, and high female representation in parliament, on increasing the national economic participation of women, in Croatia, for the year 1999, might have changed over the years, as these factors were found absent in Table 5.22.

VISUALIZING NECESSARY AND SUFFICIENT CONDITIONS

To visualize necessary and sufficient conditions, we can apply the XYplot() function from the SetMethods package, and the xy.plot() function from the QCA package (Oana and Schneider 2018; Dusa 2019). To visualize SUIN conditions, researchers can also apply the pimplot() function, from the SetMethods package (Oana and Schneider 2018).

I have applied the XYplot() function to visualize, MFISR as a necessary condition for HNWEB (in Figure 6.1), and for the enhanced intermediate solution (Figure 6.2).

TABLE 6.14 Codes for Visualizing MFISR as a Necessary Condition

```
PANELB$Observations<-NULL
PANELB$COUNTRY<-NULL
PANELB$YEAR<-NULL
PANELB$X<-NULL
XYplot(MFISR, HNWEB, data=PANELB, relation = "necessity",
  enhance = TRUE, jitter=TRUE, clabels = seq(nrow(PANELB)))
```

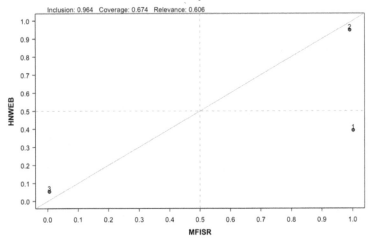

Necessity relation

Inclusion: 0.964   Coverage: 0.674   Relevance: 0.606

TABLE 6.15   Codes for Visualizing the Enhanced Intermediate Solution

```
PANELB$Observations<-NULL
PANELB$COUNTRY<-NULL
PANELB$YEAR<-NULL
PANELB$X<-NULL
XYplot(MFISR*GDPR*DEMC*DPCL*LTRCYR*NEFFECTR*FRPR*HRTR,
  HNWEB, data=PANELB, relation = "sufficiency", enhance =
  TRUE, jitter=TRUE, clabels = seq(nrow(PANELB)))
```

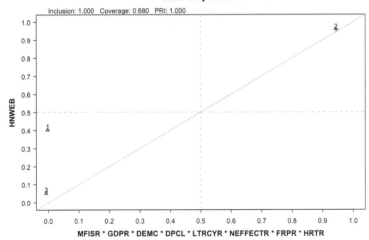

Sufficiency relation

Inclusion: 1.000   Coverage: 0.680   PRI: 1.000

TABLE 6.16 Model B (Years: 1999 and 2007)

CONCEPT	DIMENSION/ CHARACTERISTIC	INDICATOR/VARIABLE	ABBREVIATION	TYPE OF SET	TYPE OF CALIBRATION	SET LABEL
Microfinance Institutions	Presence of microfinance institutions	Presence of MFIs **Source:** World Bank| Databank| MIX Market	MFISB	Crisp()	Threshold=1	Set of countries with presence of microfinance institutions
Access to Political Freedom	The extent of democracy within the country	Democracy Scale (DEM), **Source:** Polity IV Dataset	DEMC	Fuzzy()	DEMB: Direct Calibration Method (e=−66, c=1, i=9)	Set of countries with high extent of democracy
	Access to political and civil rights	Political Rights & Civil Liberty Scale (PCL) **Source:** Freedom Rating Scale, Freedom House Reports	DPCL		PCLB: Direct Calibration Method (i=2.0, c=3.2, e=5.0)	Set of countries with high access to political and civil rights
Access to Economic Facilities	Economic growth – > increases opportunities available to individuals for production, consumption, and exchange	Gross Domestic Product **Source:** World Bank	GDPRB	Fuzzy()	Direct Calibration Method (thresholds," e= 3500000000, c= 20000000000, i= 600000000000")	Set of countries with high extent of economic growth

Access to Social Opportunities	Access to basic education and healthcare, which will help an individual live a better life	% of literate females ages 15–24 **Source:** World Bank Life expectancy at birth for females, measured in years **Source:** World Bank	LTRCYR LIFEXFR	Fuzzy()	Direct Calibration Method (thresholds= e=98.70, c=99.62, i=99.80) Direct Calibration Method (thresholds; e=76.00, c=77.00, i=79.00)	Set of countries with a high percentage of literate females Set of countries with high female life expectancy
Access to Transparency Guarantees	Society operates on the basic presumption of trust. Need for openness, guarantees of disclosure, and low corruption	Control of Corruption Scale **Source:** World Governance Indicators| World Bank	FCORB	Fuzzy()	Direct Calibration Method (thresholds, e=34.0, c=52.0, i=71.0)	Set of countries with high control of corruption
Access to Protective Security	Effectiveness of public institutions in ensuring access to basic facilities like sanitation, health, and unemployment benefits	Government Effectiveness Scale **Source:** World Governance Indicators| World Bank	FEFFECTB	Fuzzy()	Direct Calibration Method (thresholds, e=0, c=50.0, i=68.0)	Set of countries with high government effectiveness

(Continued)

TABLE 6.16 (Continued)

CONCEPT	DIMENSION/ CHARACTERISTIC	INDICATOR/VARIABLE	ABBREVIATION	TYPE OF SET	TYPE OF CALIBRATION	SET LABEL
International Human Rights Treaties (Case Study Condition)	Signed human rights treaties to eliminate all forms of racial and gender inequality	Signed 18 international human rights treaties, Source: UN Human Rights Office of the High Commissioner	HRTR	Fuzzy()	Direct Calibration method (thresholds; "e=0, c=11.5, i=14")	Set of countries with high number of international human rights treaties signed
Female Representation in Parliament (Case Study Condition)	Percentage of seats held by women in national parliaments	Proportion of seats held by women in national parliaments (%) **Source:** Millennium Development Goals, World Bank	FRPR	Fuzzy()	Direct Calibration Method (thresholds: e=0.00, c=13.00, i=20.00)	High female representation in parliament
Economic Empowerment of Women (Outcome)	Economic participation of women at the level of the national economy, for the economic sectors of own-account workers (informal economy), family workers (informal economy), employees (formal economy), employers (formal economy)	Number of women employed belonging to the age group, 15–65 years. **Source:** Employment by Sex and Status in Employment (KILM 3); ILO Modeled Estimates – Annual, calculated in thousands	Macro-condition: (HNWEB) OWNR, FAMR, EMPEB, EMPRB	FuzzyAnd()	Direct Calibration Method OWNR: (thresholds= e=5000, c=60000, i=96000) FAMR: (thresholds= e=2000, c=40000, i=68000) EMPER: (thresholds= "e=55000, c=300000, i=620000") EMPRR: (thresholds= "e=3200, c=12000, i=24000")	Set of countries with high economic participation of women

According to the assumptions of Sub-QCA Model A, to analyze how the impact of conditions have changed over time, we need to create a separate sub-model for another year (for example, 2007), apply the steps described above for that sub-model, and then compare the result from the enhanced intermediate solutions.

Sub-QCA Model B

To maintain a balance between the number of cases and conditions studied, we can also include the different years that we want to analyze (for example, years 1999 and 2007) in a single model, which I call Sub-QCA Model B. I have tabulated the calibration strategy in Table 6.16.

TESTING FOR NECESSARY CONDITIONS

To test for necessary conditions, we need to apply the same codes as Sub-QCA Model A, and I have included them in this chapter's appendix. My testing for necessary conditions shows that presence of Microfinance Institutions (MFISR) is a necessary condition that can increase the macro-economic participation of women (HNWEB), (consistency score > 0.9), Table 6.17. However, the absence of MFISR did not lead to the absence of HNWEB, as seen in Table 6.17. Hence MFISR is not a necessary condition for HNWEB.

Moreover, the relevance score (RoN) is less than the coverage score (covN), Table 6.17. Figure 6.3 also shows a skewed calibrated distribution, meaning that MFISR is a trivial necessary condition.

TABLE 6.17 Testing of MFISR as a Necessary Condition for HNWEB, Model B

	TESTING OF MFISR AS NECESSARY FOR HNWEB		
	INCLN	COVN	RON
Presence of MFIs	0.974	0.371	0.241
Absence of MFIs	0.0263	0.0501	0.8403

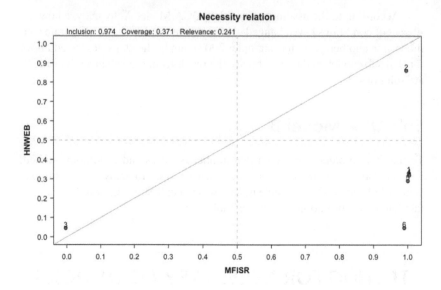

TESTING FOR SUFFICIENT CONDITIONS

To test for sufficient conditions, we need to apply the same steps as Sub-QCA Model A: create the truth tables for the presence and absence of the outcome; minimize them following Boolean Minimization; create the Standard Analysis with conservative, parsimonious, and intermediate solutions; identify and remove untenable assumptions; and create the enhanced conservative, parsimonious, and intermediate solutions. I have applied the directional expectations Table 6.4, and tabulated the enhanced intermediate solution in Table 6.18.

My enhanced intermediate solution shows that Croatia_1999 is still the most-likely case covered, but the pathways are a little bit different, as compared to the enhanced intermediate solutions from Sub-Model A (Table 6.13), and Cluster QCA (Table 5.22).

According to Table 6.18, there are two main pathways which might have increased the national economic participation of women in Croatia, for the year 1999. The first pathway shows a combination of the presence of Microfinance Institutions (MFISR), high economic growth (GDPR), and high government effectiveness (NEFFECT), but the absence of conditions like high political and civil liberties (~DPCL), high female literacy (~LTRCYR), high female life expectancy (LIFEXFR), high control of corruption (~NCORR), and high female representation in parliament (~FRPR).

TABLE 6.18   Enhanced Intermediate Solution, Model B

```
From C1P1:

M1:  MFISR*GDPR*~DPCL*~LTRCYR*~LIFEXFR*~NCORR*NEFFECTR*~FRPR -> HNWEB

                                 inclS   PRI   covS   covU  cases
------------------------------------------------------------------
1 MFISR*GDPR*~DPCL*~LTRCYR*~LIFEXFR*~NCORR* 0.999 0.998 0.360   -   HRV99
  NEFFECTR*~FRPR
------------------------------------------------------------------
                              M1 0.999 0.998 0.360

From C1P2:

M1:  MFISR*GDPR*~DEMC*~DPCL*~LTRCYR*~LIFEXFR*NEFFECTR*~FRPR -> HNWEB

                                 inclS   PRI   covS   covU  cases
------------------------------------------------------------------
1 MFISR*GDPR*~DEMC*~DPCL*~LTRCYR*~LIFEXFR* 1.000 1.000 0.358   -   HRV99
  NEFFECTR*~FRPR
------------------------------------------------------------------
                              M1 1.000 1.000 0.358
```

The second pathway shows a combination of the presence of Microfinance Institutions (MFISR), high economic growth (GDPR), and high government effectiveness (NEFFECT), but the absence of conditions like, high political and civil liberties (~DPCL), high democracy(~DEMC), high female literacy (~LTRCYR), high female life expectancy (LIFEXFR), high control of corruption (~NCORR), and high female representation in parliament (~FRPR).

Either of these pathways/causal mechanisms could have increased the national economic participation of women in Croatia, for the year 1999, and researchers are encouraged to apply post-QCA smmr() approach, case study analysis, to analyze which of these pathways worked, similar to Chapter 5.

VISUALIZATIONS

To visualize the data analysis results from the enhanced intermediate solution, Table 6.18, we can apply the XYplot() function, as demonstrated previously in Model A. I have added the figures below, and codes on Harvard Dataverse.

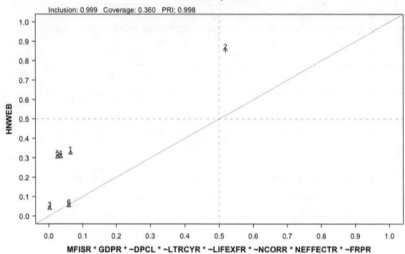

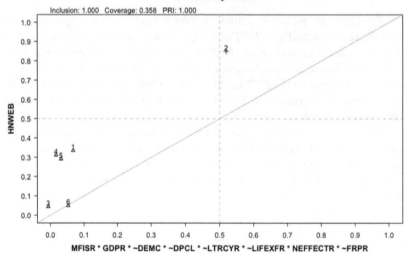

STRENGTHS AND WEAKNESSES

In this chapter, I discussed Verweij and Vis's (2021) *Sub-QCA Model* approach. The biggest strength of this approach is that it helps us understand how the condition and outcome variables have changed qualitatively over time (Ragin 1987, 2014), making this approach superior to the approach discussed in Chapter 5.

But sometimes it is difficult to maintain a balance between the number of conditions and cases studied, in which case researchers can create *macro-conditions* to reduce the number of conditions. Researchers can also apply the *two-step approach*, suggested by Schneider and Wagemann (2006), and Schneider (2019), to study remote and proximate conditions separately. I will be discussing this *"Remote-Proximate Panel"* approach in the next chapter.

However, researchers need to justify how they have selected the time-points or intervals for the years studied. Thus, the two approaches described in this chapter are more useful as a second step than as a first step in multi-method research – that is, if researchers already know the causal mechanism/s or the most-likely cases (Schneider and Rohlfing 2013).

Despite these weaknesses, comparing the results of Sub-QCA Model A, Table 6.12, and Sub-QCA Model B, Table 6.17, with Chapter 5, Table 5.22, yields several findings shared by these three models: First, Croatia is still the most-likely case. Second, although the exact pathway for sufficiency is slightly different for each model, they all show that the presence of microfinance institutions and high government effectiveness increased the economic participation of women, at the level of the national economy for Croatia, in the year 1999. Since QCA is based on equifinality, researchers should apply a post-QCA case study analysis, to understand which of these enhanced intermediate solution pathways from Tables 5.22, 6.12, and 6.17, actually increased the economic participation of women in the case of Croatia for the year 1999.

Appendix

- Dataset for Model A, Year 1999: FinalData_1999.csv
- Dataset for Model A, Year 2008: Dataset_2008.csv, available online on Harvard Dataverse

OBSERVATIONS	PRESENCEOFMFIS_2008	NUMBEROFMFIS_2008	GDP_2008
Bosnia and Herzegovina	1	18	1.91E+10
Croatia	0	0	7.03E+10
Montenegro	1	4	4.55E+09

OBSERVATIONS	DEM_2008	PCL_2008	LTRCYF_2008	LIFEXF_2008	FCOR_2008
Bosnia and Herzegovina	−66	3.5	99.84363	78.182	42.71844
Croatia	0	2	99.65321	79.6	58.25243
Montenegro	9	3	99.3606	76.755	51.94175

OBSERVATIONS	FEFFECT_2008	FPARL_2008	EUM_2008	INTHRT_2008
Bosnia and Herzegovina	33.49514	11.9	1	13
Croatia	71.35922	20.9	2	14
Montenegro	53.8835	11.1	1	11

OBSERVATIONS	OWN_2008	FAM_2008	EMPE_2008	EMPR_2008	COUNTRY	YEAR
Bosnia and Herzegovina	52019	40520	279573	13646	BiH	1999
Croatia	99998	28527	641158	23820	HRV	1999
Montenegro	5734	2394	75007	4510	MNE	1999

- Dataset for Model B, Years 1999 and 2007: FinalData_1999_2007.csv

OBSERVATIONS	PRESENCEOFMFIS	NUMBEROFMFIS	GDP	DEM	PCL
Bosnia and Herzegovina_1999	1	18	4.69E+09	−66	5
Croatia_1999	1	3	2.34E+10	0	4
Montenegro_1999	0	4	1.84E+10	0	5

OBSERVATIONS	LTRCYF	LIFEXF	FCOR	FEFFECT	FPARL	EUM	INTHRT
Bosnia and Herzegovina_1999	98.73545	76.581	NA	NA	NA	0	8
Croatia_1999	99.6044	76.55	NA	NA	7.9	0	8
Montenegro_1999	98.98921	76.11	NA	NA	NA	0	0

OBSERVATIONS	OWN	FAM	EMPE	EMPR	COUNTRY	YEAR
Bosnia and Herzegovina_1999	49876	59102	241554	10745	BiH	1999
Croatia_1999	82003	68965	574492	24269	HRV	1999
Montenegro_1999	5022	2040	55634	3255	MNE	1999

SUB-QCA MODEL A, YEAR 1999, SUPPLEMENTARY CODES

Testing for necessary conditions, presence and absence of outcome:

```
QCAfit(DATA$MFISR, DATA$HNWEB, cond.lab="Presence of
MFIs", necessity = TRUE, consH = TRUE)
QCAfit(DATA$GDPR, DATA$HNWEB, cond.lab="High GDP",
necessity = TRUE, consH = TRUE)
QCAfit(DATA$DEMC, DATA$HNWEB, cond.lab="High DEMC",
necessity = TRUE, consH = TRUE)
QCAfit(DATA$DPCL, DATA$HNWEB, cond.lab="High DPCL",
necessity = TRUE, consH = TRUE)
QCAfit(DATA$LTRCYR, DATA$HNWEB, cond.lab="High LTRCYR",
necessity = TRUE, consH = TRUE)
QCAfit(DATA$LIFEXFR, DATA$HNWEB, cond.lab="High
LIFEXFR", necessity = TRUE, consH = TRUE)
```

```
QCAfit(DATA$NCORR, DATA$HNWEB, cond.lab="High NCORR",
necessity = TRUE, consH = TRUE)
QCAfit(DATA$NEFFECTR, DATA$HNWEB, cond.lab="High
NEFFECT", necessity = TRUE, consH = TRUE)
QCAfit(DATA$FRPR, DATA$HNWEB, cond.lab="High FRPR",
necessity = TRUE, consH = TRUE)
QCAfit(DATA$HRTR, DATA$HNWEB, cond.lab="High HRTR",
necessity = TRUE, consH = TRUE)
QCAfit(1-DATA$MFISR, DATA$HNWEB, cond.lab="Absence of
MFIs", necessity = TRUE, consH = TRUE)
QCAfit(1-DATA$GDPR, DATA$HNWEB, cond.lab="Absence
ofHigh GDP", necessity = TRUE, consH = TRUE)
QCAfit(1-DATA$DEMC, DATA$HNWEB, cond.lab="Absence of
High DEMC", necessity = TRUE, consH = TRUE)
QCAfit(1-DATA$DPCL, DATA$HNWEB, cond.lab="Absence of
High DPCL", necessity = TRUE, consH = TRUE)
QCAfit(1-DATA$LTRCYR, DATA$HNWEB, cond.lab="Absence of
High LTRCYR", necessity = TRUE, consH = TRUE)
QCAfit(1-DATA$LIFEXFR, DATA$HNWEB, cond.lab="Absence of
High LIFEXFR", necessity = TRUE, consH = TRUE)
QCAfit(1-DATA$NCORR, DATA$HNWEB, cond.lab="Absence of
High NCORR", necessity = TRUE, consH = TRUE)
QCAfit(1-DATA$NEFFECTR, DATA$HNWEB, cond.lab="Absence
of High NEFFECT", necessity = TRUE, consH = TRUE)
QCAfit(1-DATA$FRPR, DATA$HNWEB, cond.lab="Absence of
High FRPR", necessity = TRUE, consH = TRUE)
QCAfit(1-DATA$HRTR, DATA$HNWEB, cond.lab="Absence of
High HRTR", necessity = TRUE, consH = TRUE)
```

OBSERVATIONS	PRESENCEOFMFIS	NUMBEROFMFIS	GDP	DEM	PCL
Bosnia and Herzegovina_1999	1	18	4.69E+09	−66	5
Croatia_1999	1	3	2.34E+10	0	4
Montenegro_1999	0	4	1.84E+10	0	5
Bosnia and Herzegovina_2007	1	18	1.58E+10	−66	3.5
Croatia_2007	1	3	6.02E+10	0	2
Montenegro_2007	1	4	3.68E+09	9	3

OBSERVATIONS	LTRCYF	LIFEXF	FCOR	FEFFECT	FPARL	EUM
Bosnia and Herzegovina_1999	98.73545	76.581	NA	NA	NA	0
Croatia_1999	99.6044	76.55	NA	NA	7.9	0
Montenegro_1999	98.98921	76.11	NA	NA	NA	0
Bosnia and Herzegovina_2007	99.84363	78.047	41.26213	20.38835	11.9	1
Croatia_2007	99.65321	79.25	58.73787	68.93204	20.9	2
Montenegro_2007	99.3606	76.484	47.08738	47.57281	11.1	0

OBSERVATIONS	INTHRT	OWN	FAM	EMPE	EMPR	COUNTRY	YEAR
Bosnia and Herzegovina_1999	8	49876	59102	241554	10745	BiH	1999
Croatia_1999	8	82003	68965	574492	24269	HRV	1999
Montenegro_1999	0	5022	2040	55634	3255	MNE	1999
Bosnia and Herzegovina_2007	12	49631	29957	250431	10485	BiH	2007
Croatia_2007	14	96374	28923	629619	22952	HRV	2007
Montenegro_2007	11	6861	2204	69702	5287	MNE	2007

SUB-QCA MODEL B, YEARS 1999 AND 2007

To check for necessary conditions and presence of outcome:

```
QCAfit(PANELC$MFISR, PANELC$HNWEB, cond.lab="Presence
of MFIs", necessity = TRUE, consH = TRUE)
QCAfit(PANELC$GDPR, PANELC$HNWEB, cond.lab="High GDP",
necessity = TRUE, consH = TRUE)
QCAfit(PANELC$DEMC, PANELC$HNWEB, cond.lab="High
DEMC", necessity = TRUE, consH = TRUE)
QCAfit(PANELC$DPCL, PANELC$HNWEB, cond.lab="High DPCL",
necessity = TRUE, consH = TRUE)
```

```
QCAfit(PANELC$LTRCYR, PANELC$HNWEB, cond.lab="High
LTRCYR", necessity = TRUE, consH = TRUE)

QCAfit(PANELC$LIFEXFR, PANELC$HNWEB, cond.lab="High
LIFEXFR", necessity = TRUE, consH = TRUE)

QCAfit(PANELC$NCORR, PANELC$HNWEB, cond.lab="High
NCORR", necessity = TRUE, consH = TRUE)

QCAfit(PANELC$NEFFECTR, PANELC$HNWEB, cond.lab="High
NEFFECT", necessity = TRUE, consH = TRUE)

QCAfit(PANELC$FRPR, PANELC$HNWEB, cond.lab="High FRPR",
necessity = TRUE, consH = TRUE)

QCAfit(PANELC$HRTR, PANELC$HNWEB, cond.lab="High
HRTR", necessity = TRUE, consH = TRUE)
```

To check for necessary conditions, absence of outcome:

```
QCAfit(1-PANELC$MFISR, PANELC$HNWEB, cond.lab="Absence
of MFIs", necessity = TRUE, consH = TRUE)

QCAfit(1-PANELC$GDPR, PANELC$HNWEB, cond.lab="Absence
ofHigh GDP", necessity = TRUE, consH = TRUE)

QCAfit(1-PANELC$DEMC, PANELC$HNWEB, cond.lab="Absence
of High DEMC", necessity = TRUE, consH = TRUE)

QCAfit(1-PANELC$DPCL, PANELC$HNWEB, cond.lab="Absence
of High DPCL", necessity = TRUE, consH = TRUE)

QCAfit(1-PANELC$LTRCYR, PANELC$HNWEB, cond.
lab="Absence of High LTRCYR", necessity = TRUE, consH
= TRUE)

QCAfit(1-PANELC$LIFEXFR, PANELC$HNWEB, cond.
lab="Absence of High LIFEXFR", necessity = TRUE, consH
= TRUE)

QCAfit(1-PANELC$NCORR, PANELC$HNWEB, cond.lab="Absence
of High NCORR", necessity = TRUE, consH = TRUE)

QCAfit(1-PANELC$NEFFECTR, PANELC$HNWEB, cond.
lab="Absence of High NEFFECT", necessity = TRUE, consH
= TRUE)

QCAfit(1-PANELC$FRPR, PANELC$HNWEB, cond.lab="Absence
of High FRPR", necessity = TRUE, consH = TRUE)

QCAfit(1-PANELC$HRTR, PANELC$HNWEB, cond.lab="Absence
of High HRTR", necessity = TRUE, consH = TRUE)
```

Creating a truth table, presence and absence of outcome:

```
ttEMPC<-truthTable(data=PANELC, outcome="HNWEB",
conditions= c("MFISR", "GDPR", "DEMC", "DPCL", "LTRCYR",
"LIFEXFR", "NCORR", "NEFFECTR", "FRPR", "HRTR"), incl.
cut = 0.85, sort.by = "incl,n", pri.cut=0.51, dcc= TRUE,
decreasing=FALSE,complete=TRUE, show.cases = TRUE)
ttEMPC
ttempc<-truthTable(data=PANELC, outcome="~HNWEB",
conditions= c("MFISR", "GDPR", "DEMC", "DPCL", "LTRCYR",
"LIFEXFR", "NCORR", "NEFFECTR", "FRPR", "HRTR"), incl.
cut = 0.85, sort.by = "incl,n", pri.cut=0.51, dcc= TRUE,
decreasing=FALSE,complete=TRUE, show.cases = TRUE)
ttempc
```

Boolean minimization of the truth table:

```
sol _ EMPC<-minimize(input=ttEMPC, details=TRUE, row.
dom=TRUE)
sol _ EMPC
sol _ empc<-minimize(input=ttempc, details=TRUE, row.
dom=TRUE)
sol _ empc
```

For the conservative, parsimonious, and intermediate solutions (presence of outcome):

```
sol _ CEMPC <- minimize(ttEMPC, details=TRUE, row.
dom=TRUE)
sol _ CEMPC
sol _ PEMPC <-minimize(ttEMPC, include="?",
details=TRUE, row.dom=TRUE)
sol _ PEMPC
sol _ IEMPC <- minimize(ttEMPC, include = "?", dir.
exp = "1, 1, 1, 1, 1, 1, 1, 1, 1, 1",details=TRUE, row.
dom=TRUE)
sol _ IEMPC
```

For the conservative, parsimonious, and intermediate solutions (absence of outcome):

```
sol _ cempc <- minimize(ttempc, details=TRUE, row.
dom=TRUE)
sol _ cempc
sol _ pempc<-minimize(ttempc, include="?", details=TRUE,
row.dom=TRUE)
sol _ pempc
sol _ iempc <- minimize(ttempc, include = "?", dir.
exp = "0, 0, 0, 0, 0, 0, 0, 0, 0, 0",details=TRUE, row.
dom=TRUE)
sol _ iempc
```

For identifying errors:

```
TYPETWOC<-findRows(obj=ttEMPC, type=2)
TYPETWOC
findRows(obj=ttEMPC, type=3)
```

For enhanced conservative, parsimonious, and intermediate solutions:

```
sol _ CEMPNC<-minimize(ttEMPC, details=TRUE,
exclude=c(TYPETWOC))
sol _ CEMPNC
sol _ PEMPNC<-minimize(ttEMPC, include="?",
details=TRUE, exclude=c(TYPETWOC))
sol _ PEMPNC
sol _ IEMPNC<-minimize(ttEMPC, include = "?", dir.
exp = "1, 1, 1, 1, 1, 1, 1, 1, 1, 1",details=TRUE,
exclude=c(TYPETWOC))
sol _ IEMPNC
```

Visualizing the necessary and enhanced intermediate solutions:

```
PANELC$Observations<-NULL

PANELC$COUNTRY<-NULL

PANELC$YEAR<-NULL

PANELC$X<-NULL

XYplot(MFISR, HNWEB, data=PANELC, relation = "neces-
sity", enhance = TRUE, jitter=TRUE, clabels =
seq(nrow(PANELC)))

XYplot("MFISR*GDPR*~DEMC*~DPCL*~LTRCYR*~LIFEXFR*NEF
FECTR*~FRPR", HNWEB, data=PANELC, relation = "suf-
ficiency", enhance = TRUE, jitter=TRUE, clabels =
seq(nrow(PANELC)))

XYplot("MFISR*GDPR*~DPCL*~LTRCYR*~LIFEXFR*~NCORR*NEF
FECTR*~FRPR", HNWEB, data=PANELC, relation = "suf-
ficiency", enhance = TRUE, jitter=TRUE, clabels =
seq(nrow(PANELC)))
```

Remote-Proximate Panel

7

Chapter outline:

INTRODUCTION

In this chapter, I will discuss Schneider and Wagemann (2006) and Schneider's (2019) two-step QCA model as a third approach that researchers can apply to analyze panel data in QCA. I have termed this approach *Remote-Proximate Panel*.

In *Remote-Proximate Panel*, researchers first need to differentiate between remote conditions and proximate conditions. According to Schneider (2019), remote conditions are contextual conditions, given to actors, and occur

DOI: 10.1201/9781003384595-7

far away in time and space, as compared to when the outcome happens. These remote conditions can also be defined as outcome-enabling conditions, as their main function is to identify the context under which the outcome occurs, but they cannot causally explain the outcome by themselves (Schneider and Wagemann 2006). By contrast, proximate conditions are conditions that can change over time due to choices made by actors, and these conditions occur closer to the outcome in terms of time and space (Schneider 2019).

To identify these remote conditions, we need to first test for necessary SUIN conditions, with the help of the superSubset() function (Schneider 2019). Before applying the superSubset() function, researchers need to apply the cluster() function, to see if the consistency and coverage scores change over the countries and years studied. If the results from the cluster() analysis show that the impact of conditions does not change over time, i.e. pooled consistency score is 1, then researchers cannot apply a two-step QCA model.

Next, researchers need to test for remote-proximate conditions as sufficient conditions (Schneider 2019). To do this, researchers need to include the remote conditions that have been identified as necessary SUIN conditions, and the remaining conditions, as sufficient-proximate ones.

I have applied my research data to demonstrate this model. For this chapter, I tested conditions like access to political freedom, access to economic facilities, access to transparency guarantees, access to social opportunities, access to protective security, and signing of international human rights treaties as remote conditions that can increase the national economic participation of women, in BiH, HRV, and MNE (Sen 2000). I have defined these factors as structural and contextual in nature, since these conditions already existed when my case studies became independent, in the year 1995 (Bhattacharya 2020).

I have included my other conditions, presence of microfinance institutions and female representation in parliament, as proximate conditions, because my case countries chose these conditions after independence (Bhattacharya 2020). Since Montenegro had not yet started its EU membership process during the years 1999–2007, I did not include EUMR as one of my conditions for this approach.

In the remaining portion of this chapter, I discuss the main assumptions of this model, test for remote and proximate conditions, and conclude by discussing the strengths and weaknesses of this model.

ASSUMPTIONS

The assumptions of this model are the same as those of cross-sectional QCA: that is, causal asymmetry, conjunctural causation, and equifinality. Along with these assumptions, we need to structure the dataset in long format, test first for remote conditions as necessary SUIN conditions, and then for SUIN remote-proximate conditions as sufficient conditions.

CALIBRATION STRATEGY APPLIED

Since the data structure for this model is different, I have tabulated my calibration strategy in Table 7.1.

Step 1: Testing for remote SUIN conditions

According to Schneider (2019), the first step in this model is to apply the cluster() function, to analyze whether the impact of conditions changes over time.

My cluster() analysis results showed that the impact of conditions does vary across countries and over years studied, as the pooled consistency score is less than 1. Therefore, I tested for remote SUIN conditions next. My testing for remote SUIN conditions showed that I have no SUIN conditions that can pass the necessary threshold (consistency score > 0.9).

If researchers cannot find any SUIN remote conditions with consistency greater than 0.9, then they need to adopt other panel data approaches that have been discussed in this book. However, for demonstration purposes, I modified my outcome variable to one of the economic sectors, "OWNR" (economic participation of women as own-account workers).

My cluster analysis results show that the pooled consistency scores for OWNR, are less than 1. My testing for remote SUIN conditions (Table 7.4) shows that there are 7 SUIN remote conditions. The next steps are to figure out whether there are any deviant consistency in kind (DCK) cases (located in the upper-left quadrant) and to test for skewness. To identify these DCK cases, I have applied the pimplot() function, seen in Table 7.5.

My pimplot() graphs show that the observation, Croatia_1999, is a deviant case (X<0.5 with Y>0.5, upper-left quadrant), for the remote SUIN conditions, *"GDPR + LIFEXFR," "GDPR + HRTR," "FCORR + LIFEXFR," and "LIFEXFR + HRTR."*

I have attached the graphs in Figures 7.1–7.7. Since there is just one case, Croatia_1999, with deviant consistency in kind (DCC) value, for some of the remote SUIN conditions, I have retained these conditions and the case. If there are multiple DCC cases, then researchers are encouraged to add/remove conditions and/or cases (Schneider 2019).

TABLE 7.1 Calibration Strategy

CONCEPT	DIMENSION/ CHARACTERISTIC	INDICATOR/ VARIABLE	ABBREVIATION	TYPE OF SET	TYPE OF CALIBRATION	SET LABEL
Presence of Microfinance Institutions	Presence of microfinance institutions	Presence of MFIs Source: World Bank| Databank| MIX Market	MFISR	Fuzzy()	Direct Calibration Method (threshold=1)	Set of countries with high presence of microfinance institutions
Access to Political Freedom	The extent of democracy within the country	Democracy Scale (DEM), Source: Polity IV Dataset	DEMR	Fuzzy()	DEMB: Direct Calibration Method (e=–67, c=5.3, i=7)	Set of countries with high extent of democracy
	Access to political and civil rights	Political Rights & Civil Liberty Scale (PCL) Source: Freedom Rating Scale, Freedom House Reports	DPCL		PCLB: Direct Calibration Method (i=1.9, c=2.6, e=5.1)	Set of countries with high access to political and civil rights
Access to Economic Facilities	Economic growth – > increases opportunities available to individuals for production, consumption, and exchange	Gross Domestic Product Source: World Bank	GDPR	Fuzzy()	Direct Calibration Method (thresholds," e= 980000000, c= 30000000000, i= 500000000000")	Set of countries with high extent of economic growth

(Continued)

TABLE 7.1 (Continued)

CONCEPT	DIMENSION/ CHARACTERISTIC	INDICATOR/ VARIABLE	ABBREVIATION	TYPE OF SET	TYPE OF CALIBRATION	SET LABEL
Access to Social Opportunities	Access to basic education and healthcare, which will help an individual live a better life	% of literate females ages 15–24 **Source:** World Bank	LTRCYF	Fuzzy()	Direct Calibration Method (thresholds= e=98.70, c=99.70, i=99.84)	Set of countries with a high percentage of literate females
		Life expectancy at birth for females, measured in years **Source:** World Bank	LIFEXFR		Direct Calibration Method (thresholds; e=76.00, c=77.85, i=79.00)	Set of countries with high female life expectancy
Access to Transparency guarantees	Society operates on the basic presumption of trust. Need for openness, guarantees of disclosure, and low corruption	Control of Corruption Scale **Source:** World Governance Indicators| World Bank	NCORR	Fuzzy()	Direct Calibration Method (thresholds, e=33.0, c=54.0, i=62.0)	Set of countries with high control of corruption
Access to Protective Security	Effectiveness of public institutions in ensuring access to basic facilities like sanitation, health, and unemployment benefits	Government Effectiveness Scale **Source:** World Governance Indicators| World Bank	NEFFECTR	Fuzzy()	Direct Calibration Method (thresholds, e=10.0, c=55.0, i=69.0)	Set of countries with high government effectiveness

International Human Rights Treaties (Case Study Condition)	Signed human rights treaties to eliminate all forms of racial and gender inequality	Signed 18 international human rights treaties Source: UN Human Rights Office of the High Commissioner	HRTR	Fuzzy()	Ragin's Direct Calibration: (thresholds; e=0, c=10.5, l=14)	Set of countries with high number of international human rights treaties signed
Female Representation in Parliament (Case Study Condition)	Percentage of seats held by women in national parliaments	Proportion of seats held by women in national parliaments (%) **Source:** Millennium Development Goals, World Bank	FRPR	Fuzzy()	Direct Calibration Method (thresholds: e=7.00, c=17.00, i=21.00)	High female representation in parliament
Economic Development of Women (Outcome)	Economic participation of women at the level of the national economy, for the economic sectors of own-account workers (informal economy), family workers (informal economy), employees (formal economy), employers (formal economy)	Number of women employed belonging to the age group, 15–65 years. **Source:** Employment by Sex and Status in Employment (KILM 3); ILO Modeled Estimates – Annual, calculated in thousands	Macro-condition: (HNWE) OWNR, FAMR, EMPER, EMPRR ---- Alternative macro-condition created for this approach: FE (Formal Economy): EMPER, EMPRR INFE (Informal Economy): OWNR, FAMR	FuzzyAnd()	Direct Calibration Method OWNR: (thresholds= e=4000, c=77000, i=110000) FAMR: (thresholds= e=1700, c=43000, i=60000) EMPER: (thresholds= " e=55000, c=300000, i=610000") EMPRR: (thresholds= " e=3000, c=15000, i=25000")	Set of countries with high economic participation of women

The next step is to check the *skewness* of these remote SUIN conditions, as seen in Table 7.6. To do this, we need to create new remote SUIN conditions, with fuzzyor(), and then apply the skew.check() option. The results from the skewness tests, Table 7.7, show that there is a high percentage of skewness in the remote SUIN conditions (greater than 20%). In such cases, researchers can choose to recalibrate their original data, as these remote SUIN conditions might be trivial necessary conditions (Schneider 2019). For demonstration purposes, I have continued with the next step: testing for sufficient (remote-proximate) conditions.

TABLE 7.2   Codes for Cluster Analysis

```
cluster(data = TWO_STEP_PANEL, results = "MFISR", outcome
= "OWNR",unit_id = "COUNTRY", cluster_id = "YEAR",
necessity=TRUE, wicons=TRUE)
cluster(data = TWO_STEP_PANEL, results = "GDPR", outcome
= "OWNR",unit_id = "COUNTRY", cluster_id = "YEAR",
necessity=TRUE, wicons=TRUE)
cluster(data = TWO_STEP_PANEL, results = "DEMC", outcome
= "OWNR",unit_id = "COUNTRY", cluster_id = "YEAR",
necessity=TRUE, wicons=TRUE)
cluster(data = TWO_STEP_PANEL, results = "DPCL", outcome
= "OWNR",unit_id = "COUNTRY", cluster_id = "YEAR",
necessity=TRUE, wicons=TRUE)
cluster(data = TWO_STEP_PANEL, results = "LTRCYR",
outcome = "OWNR",unit_id = "COUNTRY", cluster_id =
"YEAR", necessity=TRUE, wicons=TRUE)
cluster(data = TWO_STEP_PANEL, results = "LIFEXFR",
outcome = "OWNR",unit_id = "COUNTRY", cluster_id =
"YEAR", necessity=TRUE, wicons=TRUE)
cluster(data = TWO_STEP_PANEL, results = "NCORR", outcome
= "OWNR",unit_id = "COUNTRY", cluster_id = "YEAR",
necessity=TRUE, wicons=TRUE)
cluster(data = TWO_STEP_PANEL, results = "NEFFECTR",
outcome = "OWNR",unit_id = "COUNTRY", cluster_id =
"YEAR", necessity=TRUE, wicons=TRUE)
cluster(data = TWO_STEP_PANEL, results = "FRPR", outcome
= "OWNR",unit_id = "COUNTRY", cluster_id = "YEAR",
necessity=TRUE, wicons=TRUE)
cluster(data = TWO_STEP_PANEL, results = "HRTR", outcome
= "OWNR",unit_id = "COUNTRY", cluster_id = "YEAR",
necessity=TRUE, wicons=TRUE)
```

I have tabulated the codes for cluster analysis, in Table 7.2, and for SUIN remote conditions in Table 7.3.

TABLE 7.3 Codes for Testing SUIN Remote Conditions, with OWNR as Outcome Variables

```
SUIN_A<-superSubset(data=TWO_STEP_PANEL,
 conditions=c("GDPR", "DEMC", "DPCL", "NCORR", "NEFFECTR",
 "LTRCYR", "LIFEXFR", "HRTR"), outcome="OWNR", relation =
 "necessity", incl.cut = 0.9, cov.cut=0.6, ron.cut=0.5)
```

TABLE 7.4 Data Analysis Results from the Testing of SUIN Remote Conditions, with OWNR

		inclN	RoN	covN
1	GDPR + NCORR	0.904	0.769	0.657
2	GDPR + LIFEXFR	0.906	0.869	0.772
3	GDPR + HRTR	0.900	0.735	0.624
4	DPCL + NEFFECTR	0.925	0.698	0.602
5	NCORR + LIFEXFR	0.922	0.702	0.604
6	NEFFECTR + LIFEXFR	0.970	0.742	0.655
7	LIFEXFR + HRTR	0.930	0.717	0.620

TABLE 7.5 Codes for Identifying Deviant Consistency in Kind Cases

```
pimplot(data=TWO_STEP_PANEL, results=SUIN_A,
 outcome="OWNR", necessity=TRUE, jitter=TRUE, all_labels =
 TRUE)
```

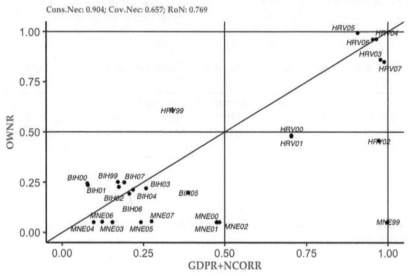

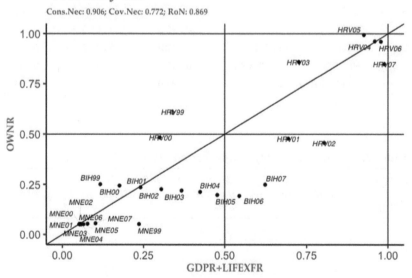

FIGURES 7.1–7.7 Graphs from the testing of SUIN Necessary conditions

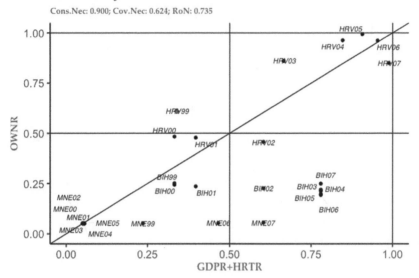

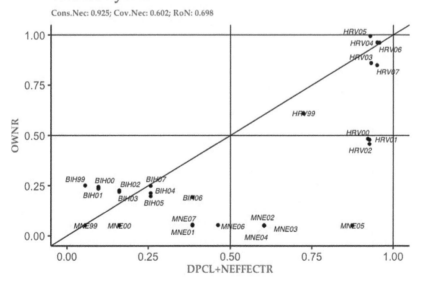

FIGURES 7.1–7.7 (*Continued*)

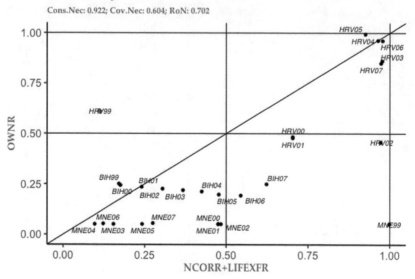

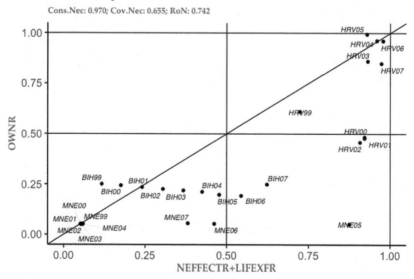

FIGURES 7.1–7.7 (*Continued*)

Necessity Plot

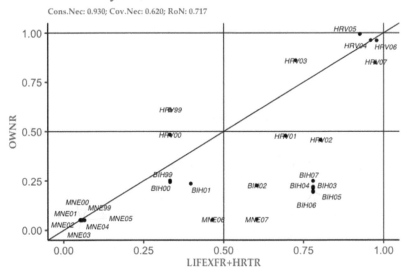

Cons.Nec: 0.930; Cov.Nec: 0.620; RoN: 0.717

FIGURES 7.1–7.7 *(Continued)*

TABLE 7.6   Codes for Skewness Tests

```
TWO_STEP_PANEL$GDPLIFE<-fuzzyor(TWO_STEP_PANEL$GDPR,
 TWO_STEP_PANEL$LIFEXFR)
TWO_STEP_PANEL$GDPHRT<-fuzzyor(TWO_STEP_PANEL$GDPR,
 TWO_STEP_PANEL$HRTR)
TWO_STEP_PANEL$DPCLNEFFECT<-fuzzyor(TWO_STEP_PANEL$DPCL,
 TWO_STEP_PANEL$NEFFECTR)
TWO_STEP_PANEL$NCORNEFFECT<-fuzzyor(TWO_STEP_PANEL$NCORR,
 TWO_STEP_PANEL$NEFFECTR)
TWO_STEP_PANEL$NCORLIFE<-fuzzyor(TWO_STEP_PANEL$NCORR,
 TWO_STEP_PANEL$LIFEXFR)
TWO_STEP_PANEL$NEFFECTLIFE<-fuzzyor(TWO_STEP_
 PANEL$NEFFECTR, TWO_STEP_PANEL$LIFEXFR)
TWO_STEP_PANEL$LIFEHRT<-fuzzyor(TWO_STEP_PANEL$LIFEXFR,
 TWO_STEP_PANEL$HRTR)
skew.check(TWO_STEP_PANEL$GDPLIFE)
skew.check(TWO_STEP_PANEL$GDPHRT)
skew.check(TWO_STEP_PANEL$DPCLNEFFECT)
skew.check(TWO_STEP_PANEL$NCORNEFFECT)
skew.check(TWO_STEP_PANEL$NCORLIFE)
skew.check(TWO_STEP_PANEL$NEFFECTLIFE)
skew.check(TWO_STEP_PANEL$LIFEHRT)
```

TABLE 7.7   Results from the Skewness Tests

```
[1] "Cases > 0.5 / Total number of cases: 9 / 27 = 33.33 %"
[1] "Cases > 0.5 / Total number of cases: 13 / 27 = 48.15 %"
[1] "Cases > 0.5 / Total number of cases: 13 / 27 = 48.15 %"
[1] "Cases > 0.5 / Total number of cases: 11 / 27 = 40.74 %"
[1] "Cases > 0.5 / Total number of cases: 11 / 27 = 40.74 %"
[1] "Cases > 0.5 / Total number of cases: 12 / 27 = 44.44 %"
[1] "Cases > 0.5 / Total number of cases: 14 / 27 = 51.85 %"
```

Step 2: Testing for remote-proximate conditions

The next step is to test for sufficient conditions, by testing the remote SUIN conditions, and the remaining proximate conditions in one truth table. There are two ways of adding remote SUIN conditions (Schneider 2019): separately, or as macro-conditions. I have demonstrated both these ways below.

FIRST WAY: ADDING REMOTE SUIN CONDITIONS SEPARATELY

The first approach is to add remote SUIN conditions separately, by adding these SUIN conditions as a part of the truthTable () function, Table 7.8.

TABLE 7.8   Codes for Creating the Truth Table, with remote SUIN Conditions Separately

```
# Presence of Outcome, OWNR:
ttOWND<-truthTable(data=TWO_STEP_PANEL, outcome="OWNR",
 conditions= c("MFISR", "GDPR", "DEMC", "DPCL", "LTRCYR",
 "LIFEXFR", "NCORR", "NEFFECTR", "FRPR", "HRTR"), incl.cut
 = 0.85, sort.by = "incl,n", pri.cut=0.51, dcc= TRUE,
 decreasing=FALSE,complete=TRUE, show.cases = TRUE)
ttOWND
# Absence of Outcome, ~OWNR:
ttownd<-truthTable(data=TWO_STEP_PANEL, outcome="~OWNR",
 conditions= c("MFISR", "GDPR", "DEMC", "DPCL", "LTRCYR",
 "LIFEXFR", "NCORR", "NEFFECTR", "FRPR", "HRTR"), incl.cut
 = 0.85, sort.by = "incl,n", pri.cut=0.51, dcc= TRUE,
 decreasing=FALSE,complete=TRUE, show.cases = TRUE)
ttownd
```

The truth table result for the presence of the outcome shows that there are three deviant consistencies in kind (dcc) cases, Table 7.9, (HRV01,HRV02,HRV00, with OUT=1).

In such scenarios, researchers can increase or decrease the incl.cut() score to change the outcome of these pathways to zero. For now, I have left these cases in the truth table since they are just three out of 27 cases. The next step is to analyze the truth table for the absence of the outcome, to ensure that there are no issues related to simultaneous subset relations.

The remaining steps are similar to the testing of sufficient conditions in the previous chapters. These steps include minimizing the truth tables, creating conservative, parsimonious, and intermediate solutions, for the presence as well as absence of the outcome. After creating the conservative, parsimonious, and intermediate solutions, we need to identify and remove the untenable assumptions, to create the enhanced standard analysis solutions (Oana, Schneider, and Thomann 2021, 130).

TABLE 7.9 Truth Table for Presence of the Outcome

OUT: output value
  n: number of cases in configuration
incl: sufficiency inclusion score
 PRI: proportional reduction in inconsistency
DCC: deviant cases consistency

	MFISR	GDPR	DEMC	DPCL	LTRCYR	LIFEXFR	NCORR	NEFFECTR	FRPR	HRTR	OUT	n	incl	PRI
864	1	1	0	1	0	1	1	1	1	1	1	5	1.000	1.000
517	1	0	0	0	0	0	0	1	0	0	1	1	1.000	1.000
607	1	0	0	1	0	1	1	1	1	0	1	1	0.973	0.879
608	1	0	0	1	0	1	1	1	1	1	1	1	0.963	0.841
591	1	0	0	1	0	0	1	1	1	0	1	1	0.963	0.747
513	1	0	0	0	0	0	0	0	0	0	0	1	0.740	0.056
545	1	0	0	0	1	0	0	0	0	0	0	2	0.699	0.023
562	1	0	0	0	1	1	0	0	0	1	0	2	0.600	0.017
641	1	0	1	0	0	0	0	0	0	0	0	1	0.571	0.029
546	1	0	0	0	1	0	0	0	0	1	0	4	0.551	0.012
642	1	0	1	0	0	0	0	0	0	1	0	1	0.533	0.025
9	0	0	0	0	0	0	1	0	0	0	0	1	0.329	0.000
197	0	0	1	1	0	0	0	1	0	0	0	1	0.252	0.000
129	0	0	1	0	0	0	0	0	0	0	0	2	0.131	0.000
193	0	0	1	1	0	0	0	0	0	0	0	3	0.129	0.000

	DCC
864	
517	
607	HRV01
608	HRV02
591	HRV00
513	BIH99
545	BIH00,BIH01
562	BIH06,BIH07
641	MNE06
546	BIH02,BIH03,BIH04,BIH05
642	MNE07
9	MNE99
197	MNE05
129	MNE00,MNE01
193	MNE02,MNE03,MNE04

Since, the first step in this approach is to test for remote SUIN conditions, to create the enhanced intermediate solutions, researchers need to look through the truth table and remove rows with *contradictory claims of necessity*, as well as rows with *contradictory simplifying assumptions* and *impossible remainders* (Oana, Schneider, and Thomann 2021, 132–139).

Rows with *contradictory claims of necessity* are rows which have no SUIN necessary conditions but do have an outcome of 1 (Oana, Schneider, and Thomann 2021, 133). Since I did not find any rows with contradictory claims of necessity in my truth table (Table 7.9), I have not included the code for identifying contradictory claims of necessity here. But, researchers are welcome to refer to Oana, Schneider, and Thomann (2021), for further detail.

We also need to remove rows with *contradictory simplifying assumptions* and *impossible remainders*. As seen below, I have applied the findRows() function, to identify rows with contradictory simplifying assumptions, and found type two errors. I then created enhanced conservative, parsimonious, and intermediate solutions.

I have tabulated these codes, in Tables 7.11 to 7.15, along with the directional expectations for the intermediate and enhanced intermediate solutions, in Table 7.10, and the results from the enhanced intermediate solution, in Table 7.16.

TABLE 7.10   Directional Expectations Between Conditions and Outcome

CONDITIONS	INCREASES ECO. PARTICIPATION	DECREASES ECO. PARTICIPATION
MFISR	Present(1)	Absent(0)
HRTR	Present(1)	Absent(0)
GDPR	Present(1)	Absent(0)
DEMC	Present(1)	Absent(0)
DPCL	Present(1)	Absent(0)
LTRCYR	Present(1)	Absent(0)
LIFEXFR	Present(1)	Absent(0)
FCORR	Present(1)	Absent(0)
NEFFECTR	Present(1)	Absent(0)
FRPR	Present(1)	Absent(0)

TABLE 7.11   Codes for Minimizing the Truth Tables

```
sol _ OWNE<-minimize(input=ttOWNRE, details=TRUE, row.dom=TRUE)
sol _ OWNE
sol _ owne<-minimize(input=ttownre, details=TRUE, row.dom=TRUE)
sol _ owne
```

TABLE 7.12   Codes for Creating the Conservative, Parsimonious, and Intermediate Solutions, Presence of the Outcome

```
sol_CONSE <- minimize(ttOWNRE, details=TRUE, row.dom=TRUE)
sol_CONSE
sol_PARSE<-minimize(ttOWNRE, include="?", details=TRUE,
 row.dom=TRUE)
sol_PARSE
sol_INTSE<- minimize(ttOWNRE, include = "?", dir.exp = "1,
 1, 1, 1, 1, 1, 1, 1, 1",details=TRUE, row.dom = TRUE)
sol_INTSE
```

TABLE 7.13   Codes for Creating the Conservative, Parsimonious, and Intermediate Solutions, Absence of the Outcome

```
sol_ownc <- minimize(ttownd, details=TRUE, row.dom=TRUE)
sol_ownc
sol_ownp<-minimize(ttownd, include="?", details=TRUE,
 row.dom=TRUE)
sol_ownp
sol_owni <- minimize(ttownd, include = "?", dir.exp = "0,
 0, 0, 0, 0, 0, 0, 0, 0, 0",details=TRUE, row.dom=TRUE)
sol_owni
```

TABLE 7.14   Codes for Identifying Untenable Assumptions

```
TYPETWO<-findRows(obj=ttOWND, type=2)
TYPETWO
findRows(obj=ttOWND, type=3)
```

TABLE 7.15   Codes for Creating the Enhanced Conservative, Parsimonious, and Intermediate Solutions

```
sol_COWNC<- minimize(ttOWND, details=TRUE, row.dom = TRUE,
 exclude=c(TYPETWO))
sol_COWNC
sol_POWNP <- minimize(ttOWND, include="?", details=TRUE,
 row.dom = TRUE, exclude=c(TYPETWO))
sol_POWNP
sol_IOWNI <- minimize(ttOWND, include="?", details=TRUE,
 dir.exp = "1, 1, 1, 1, 1, 1, 1, 1, 1, 1",
 exclude=c(TYPETWO))
sol_IOWNI
```

TABLE 7.16 The Enhanced Intermediate Solution

```
From C1P1:

M1:   MFISR*~DEMC*~DPCL*NEFFECTR + MFISR*~DEMC*NCORR*NEFFECTR*FRPR -> OWNR

                                    inclS  PRI    covS   covU
    -------------------------------------------------------------
    1        MFISR*~DEMC*~DPCL*NEFFECTR  1.000  1.000  0.265  0.072
    2  MFISR*~DEMC*NCORR*NEFFECTR*FRPR  0.954  0.908  0.551  0.358
    -------------------------------------------------------------
                                 M1  0.959  0.914  0.623

                                    cases
    -------------------------------------------------------------
    1        MFISR*~DEMC*~DPCL*NEFFECTR  HRV99
    2  MFISR*~DEMC*NCORR*NEFFECTR*FRPR  HRV00; HRV01; HRV02; HRV03,HRV04,HRV05,HRV06,HRV07
    -------------------------------------------------------------


From C1P2:

M1:   MFISR*~DEMC*DPCL*NCORR*NEFFECTR*FRPR + MFISR*~GDPR*~DEMC*~LTRCYR*~LIFEXFR*N
      EFFECTR*~FRPR*~HRTR -> OWNR

                                                inclS  PRI    covS   covU
    -------------------------------------------------------------------------
    1            MFISR*~DEMC*DPCL*NCORR*NEFFECTR*FRPR  0.954  0.908  0.551  0.466
    2  MFISR*~GDPR*~DEMC*~LTRCYR*~LIFEXFR*NEFFECTR*~FRPR*~HRTR  1.000  1.000  0.149  0.063
    -------------------------------------------------------------------------
                                           M1  0.959  0.914  0.615

                                                cases
    -------------------------------------------------------------------------
    1            MFISR*~DEMC*DPCL*NCORR*NEFFECTR*FRPR  HRV00; HRV01; HRV02; HRV03,
                                                       HRV04,HRV05,HRV06,HRV07
    2  MFISR*~GDPR*~DEMC*~LTRCYR*~LIFEXFR*NEFFECTR*~FRPR*~HRTR  HRV99
    -------------------------------------------------------------------------
```

SECOND WAY: ADDING REMOTE SUIN CONDITIONS AS MACRO-CONDITIONS

The second approach is to add these SUIN conditions as macro-conditions, shown in Table 7.17, and then test for sufficiency (Schneider 2019). Since the rest of the steps are similar to the first approach discussed above, I have tabulated only the results from the enhanced intermediate results, in Table 7.18 below, and included the codes in my appendix material, at the end of this chapter.

VISUALIZING THE ENHANCED INTERMEDIATE SOLUTIONS

To visualize the enhanced intermediate solutions, for both above-mentioned approaches, we can apply the pimplot() or XYplot() functions. I have applied the XYplot() function, codes in Table 7.19, and figures in 7.8–7.9.

TABLE 7.17   Codes for Creating the Truth Table with Macro-Conditions

```
ttOWNRE<-truthTable(TWO_STEP_PANEL, conditions= "MFISR,
GDPLIFE, GDPHRT, DPCLNEFFECT, NCORNEFFECT, NCORLIFE,
NEFFECTLIFE, LIFEHRT, FRPR",outcome="OWNR", incl.cut =
0.85, sort.by = "incl,n", pri.cut=0.51, dcc= TRUE,
decreasing=FALSE,complete=TRUE, wicons=TRUE, show.cases =
TRUE)
ttOWNRE

ttownre<-truthTable(TWO_STEP_PANEL, conditions= "MFISR,
GDPLIFE, GDPHRT, DPCLNEFFECT, NCORNEFFECT, NCORLIFE,
NEFFECTLIFE, LIFEHRT, FRPR",outcome="~OWNR", incl.cut =
0.85,
sort.by = "incl,n", pri.cut=0.51, dcc= TRUE,
decreasing=FALSE,complete=TRUE, wicons=TRUE, show.cases =
TRUE)
ttownre
```

TABLE 7.18   Enhanced Intermediate Solution

From C1P1, C1P2:
M1: MFISR*GDPLIFE*DPCLNEFFECT*NCORNEFFECT*NCORLIFE*NEFFECTLIFE*LIFEHRT*FRPR -> OWNR

	inclS	PRI	covS	covU
1 MFISR*GDPLIFE*DPCLNEFFECT*NCORNEFFECT* NCORLIFE*NEFFECTLIFE*LIFEHRT*FRPR	0.894	0.833	0.747	-
M1	0.894	0.833	0.747	

	cases
1 MFISR*GDPLIFE*DPCLNEFFECT*NCORNEFFECT* NCORLIFE*NEFFECTLIFE*LIFEHRT*FRPR	HRV01; HRV02,HRV03,HRV04,HRV05, HRV06,HRV07

TABLE 7.19   Visualizing the Enhanced Intermediate Solutions

```
TWO_STEP_PANEL$Observations<-NULL
TWO_STEP_PANEL$COUNTRY<-NULL
TWO_STEP_PANEL$YEAR<-NULL
TWO_STEP_PANEL$X<-NULL
XYplot("MFISR*~DEMC*NCORR*NEFFECTR*FRPR", OWNR, data=TWO_
STEP_PANEL, relation = "sufficiency", enhance = TRUE,
jitter=TRUE, clabels = seq(nrow(TWO_STEP_PANEL)))
XYplot("MFISR*GDPLIFE*DPCLNEFFECT*NCORNEFFECT*NCORLIFE*NE
FFECTLIFE*LIFEHRT*FRPR", OWNR, data=TWO_STEP_PANEL,
relation = "sufficiency", enhance = TRUE, jitter=TRUE,
clabels = seq(nrow(TWO_STEP_PANEL)))
```

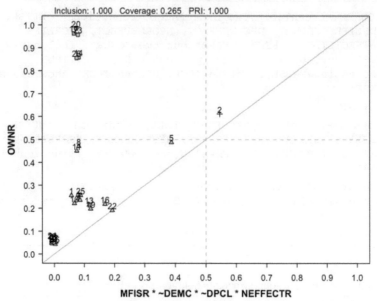

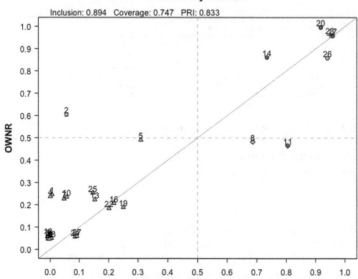

FIGURES 7.8–7.9: Enhanced Intermediate Solutions

COMPARING THE TWO APPROACHES OF ADDING REMOTE SUIN CONDITIONS

My enhanced intermediate solutions (Tables 7.16 and 7.18) show that the results from both these approaches are quite similar to one another, but not entirely.

For the first approach, Table 7.16, my typical cases were from Croatia during the years 1999–2007. For the second approach, Table 7.18, my typical cases were also from Croatia, only for the years 2001–2007.

The second approach (Table 7.18), also identified a lower number of causal conjuncts and conjunctions, which might be due to the presence of macro-conditions (Schneider 2019; Oana, Schneider, and Thomann 2021, 184).

But the cases from Croatia, 2000–2003, did turn out to be typical cases for both these approaches. The pathways from both these approaches, though similar, are a little bit different, especially with the presence of GDP as a macro-condition, in the second approach, according to Table 7.18. This might be due to the fact that the macro-conditions were a little bit *skewed,* so researchers should apply set-theoretic multi-method research, smmr(), to further analyze the causal pathways/mechanisms, from Table 7.16 as well as Table 7.18.

STRENGTHS AND WEAKNESSES

In this chapter, I demonstrated the *Remote-Proximate Panel*, first proposed by Schneider and Wagemann (2006), and later modified by Schneider (2019), as a third approach to analyzing panel data in QCA.

One of the most important advantages of this approach is that it helps us analyze panel data, while also applying the assumptions of QCA, by analyzing "qualitative change over time" (Ragin 1987, 2014), as compared to *Cluster QCA* (Chapter 5).

Compared to the Multiple Sub-QCA models discussed in Chapter 6, this model can help us reduce the number of conditions studied, by dividing the conditions into remote and proximate.

But there are a few weaknesses. First, researchers need to theoretically and/or contextually justify which conditions are remote and which are proximate to apply this model (Schneider 2019, 1123).

Second, the remote conditions need to pass the test for necessary SUIN conditions, with a consistency score of at least 0.9, coverage score of at least 0.6, and Relevance of Necessity (RoN) score of at least 0.5 (Schneider 2019, 1120).

If we cannot find any necessary condition with these thresholds, then we cannot apply this approach.

Third, we also need to be sure that the conditions included in the model do change over time and across cases studied (Schneider 2019), which we can do by first applying the cluster() function as discussed in Chapter 5 to see if the impact of conditions on the outcome changes over the cases and years studied, i.e., Pooled Consistency score (POCON) will be less than 1.0, and then applying the two-step approach (Schneider and Wagemann 2006; Schneider 2019).

Finally, since I analyzed a different outcome variable, "economic participation of women as own-account workers," OWNR, I have not compared the causal pathways from this approach with those in Chapters 5 and 6.

Appendix

* **Dataset used: Final_Panel_data.csv**

OBSERVATIONS	PRESENCE-OFMFIS	NUMBER-OFMFIS	GDP	DEM	PCL
Bosnia and Herzegovina_1999	1	18	4.69E+09	−66	5
Croatia_1999	1	3	2.34E+10	0	4
Montenegro_1999	0	4	1.84E+10	0	5
Bosnia and Herzegovina_2000	1	18	5.51E+09	−66	4.5
Croatia_2000	1	3	2.17E+10	0	2.5
Montenegro_2000	0	4	9.84E+08	7	4
Bosnia and Herzegovina_2001	1	18	5.75E+09	−66	4.5
Croatia_2001	1	3	2.32E+10	0	2
Montenegro_2001	0	4	1.16E+09	7	3
Bosnia and Herzegovina_2002	1	18	6.65E+09	−66	4
Croatia_2002	1	3	2.68E+10	0	2
Montenegro_2002	0	4	1.28E+09	7	2.5
Bosnia and Herzegovina_2003	1	18	8.37E+09	−66	4
Croatia_2003	1	3	3.47E+10	0	2
Montenegro_2003	0	4	1.71E+09	6	2.5
Bosnia and Herzegovina_2004	1	18	1.06E+10	−66	3.5
Croatia_2004	1	3	4.16E+10	0	2
Montenegro_2004	0	4	2.07E+09	6	2.5
Bosnia and Herzegovina_2005	1	18	1.12E+10	−66	3.5
Croatia_2005	1	3	4.54E+10	0	2
Montenegro_2005	0	4	2.26E+09	6	2.5
Bosnia and Herzegovina_2006	1	18	1.29E+10	−66	3
Croatia_2006	1	3	5.05E+10	0	2
Montenegro_2006	1	4	2.72E+09	9	3
Bosnia and Herzegovina_2007	1	18	1.58E+10	−66	3.5
Croatia_2007	1	3	6.02E+10	0	2
Montenegro_2007	1	4	3.68E+09	9	3

(Continued)

OBSERVATIONS	LTRCYF	LIFEXF	FCOR	FEFFECT	FPARL	EUM
Bosnia and Herzegovina_1999	98.73545	76.581	NA	NA	NA	0
Croatia_1999	99.6044	76.55	NA	NA	7.9	0
Montenegro_1999	98.98921	76.11	NA	NA	NA	0
Bosnia and Herzegovina_2000	99.84363	76.881	33.50254	17.94872	NA	0
Croatia_2000	99.6044	76.68	56.34518	66.66666	20.5	0
Montenegro_2000	98.98921	76.035	53.29949	NA	NA	0
Bosnia and Herzegovina_2001	99.84363	77.128	NA	NA	7.1	0
Croatia_2001	99.65321	78.17	NA	NA	20.5	0
Montenegro_2001	98.98921	76.022	NA	NA	NA	0
Bosnia and Herzegovina_2002	99.84363	77.333	42.92929	13.77551	NA	0
Croatia_2002	99.65321	78.4	63.63636	65.81633	20.5	0
Montenegro_2002	98.98921	76.04	53.53535	NA	NA	0
Bosnia and Herzegovina_2003	99.84363	77.508	46.46465	22.95918	16.7	1
Croatia_2003	99.65321	78.23	64.14141	67.34694	NA	1
Montenegro_2003	99.3606	76.068	41.91919	NA	NA	0
Bosnia and Herzegovina_2004	99.84363	77.659	44.87805	29.55665	16.7	1
Croatia_2004	99.65321	79.08	62.92683	68.96552	21.7	2
Montenegro_2004	99.3606	76.106	38.04878	NA	NA	0
Bosnia and Herzegovina_2005	99.84363	77.793	50.73171	25.98039	16.7	1
Croatia_2005	99.65321	78.83	58.53659	67.15686	21.7	2
Montenegro_2005	99.3606	76.172	45.85366	64.21568	NA	0
Bosnia and Herzegovina_2006	99.84363	77.92	44.39024	32.19512	14.3	1
Croatia_2006	99.65321	79.33	59.5122	69.7561	21.7	2
Montenegro_2006	99.3606	76.291	40	52.68293	8.6	0
Bosnia and Herzegovina_2007	99.84363	78.047	41.26213	20.38835	11.9	1
Croatia_2007	99.65321	79.25	58.73787	68.93204	20.9	2
Montenegro_2007	99.3606	76.484	47.08738	47.57281	11.1	0

OBSERVATIONS	INTHRT	OWN	FAM	EMPE	EMPR	COUN-TRY	YEAR
Bosnia and Herzegovina_1999	8	49876	59102	241554	10745	BiH	1999
Croatia_1999	8	82003	68965	574492	24269	HRV	1999
Montenegro_1999	0	5022	2040	55634	3255	MNE	1999
Bosnia and Herzegovina_2000	8	48935	54411	243877	10776	BiH	2000
Croatia_2000	8	75374	60513	561296	22884	HRV	2000
Montenegro_2000	0	4896	2042	55602	3281	MNE	2000
Bosnia and Herzegovina_2001	9	47803	50548	244760	10757	BiH	2001
Croatia_2001	9	74852	55749	563222	17896	HRV	2001
Montenegro_2001	0	4820	1976	55812	3331	MNE	2001
Bosnia and Herzegovina_2002	11	46485	47317	244288	10685	BiH	2002
Croatia_2002	11	72761	54332	556225	26295	HRV	2002
Montenegro_2002	0	4727	1919	55936	3371	MNE	2002
Bosnia and Herzegovina_2003	12	45518	44287	245223	10681	BiH	2003
Croatia_2003	11	97368	48321	559658	20141	HRV	2003
Montenegro_2003	0	4650	1865	56292	3421	MNE	2003
Bosnia and Herzegovina_2004	12	44543	41665	245829	10665	BiH	2004
Croatia_2004	11	113331	39525	578198	21062	HRV	2004
Montenegro_2004	0	4600	1811	56861	3450	MNE	2004
Bosnia and Herzegovina_2005	12	42294	37994	239891	10302	BiH	2005
Croatia_2005	12	134432	30940	582765	21949	HRV	2005
Montenegro_2005	0	4569	1750	57521	3469	MNE	2005
Bosnia and Herzegovina_2006	12	41522	35842	240758	10308	BiH	2006
Croatia_2006	12	112857	28208	614741	24806	HRV	2006
Montenegro_2006	10	5713	2010	64231	4370	MNE	2006
Bosnia and Herzegovina_2007	12	49631	29957	250431	10485	BiH	2007
Croatia_2007	14	96374	28923	629619	22952	HRV	2007
Montenegro_2007	11	6861	2204	69702	5287	MNE	2007

Boolean minimization for truth tables, second approach:

```
sol _ OWNE<-minimize(input=ttOWNRE, details=TRUE, row.
dom=TRUE)
sol _ OWNE
sol _ owne<-minimize(input=ttownre, details=TRUE, row.
dom=TRUE)
sol _ owne
```

For creating the conservative, parsimonious, and intermediate solutions, presence

```
sol _ CONSE <- minimize(ttOWNRE, details=TRUE, row.dom=TRUE)
sol _ CONSE
sol _ PARSE<-minimize(ttOWNRE, include="?", details=TRUE,
row.dom=TRUE)
sol _ PARSE
sol _ INTSE<- minimize(ttOWNRE, include = "?", dir.exp =
"1, 1, 1, 1, 1, 1, 1, 1, 1",details=TRUE, row.dom = TRUE)
sol _ INTSE
```

For creating the conservative, parsimonious, and intermediate solutions, absence

```
sol _ conse <- minimize(ttownre, details=TRUE, row.dom=TRUE)
sol _ conse
sol _ parse<-minimize(ttownre, include="?", details=TRUE,
row.dom=TRUE)
sol _ parse
sol _ intse<- minimize(ttownre, include = "?", dir.exp =
"1, 1, 1, 1, 1, 1, 1, 1, 1",details=TRUE, row.dom = TRUE)
sol _ intse
```

For identifying errors:

```
TYPE _ TWOB<-findRows(obj=ttOWNRE, type=2)
TYPE _ TWOB
findRows(obj=ttOWNRE, type=3)
```

For creating the enhanced conservative, parsimonious, and intermediate solutions:

```
sol_CS<-minimize(ttOWNRE, details=TRUE, exclude=c
(TYPE_TWOB))

sol _ CS

sol _ PS<-minimize(ttOWNRE, include="?", details=TRUE,
exclude=c(TYPE _ TWOB))

sol _ PS

sol _ IS<-minimize(ttOWNRE, include = "?", dir.
exp = "1, 1, 1, 1, 1, 1, 1,1,1",details=TRUE,
exclude=c(TYPE _ TWOB))

sol _ IS
```

Relevant Variation Panel

8

Chapter outline:

- Introduction
- Assumptions
- Calibration process
- Testing for necessary conditions
- Testing for sufficient conditions
- Interpreting the enhanced intermediate solution
- Visualizing the pathways
- Strengths and weaknesses

INTRODUCTION

In this chapter, I will demonstrate the fourth approach that can help us analyze panel data in QCA, if researchers are not able to decide the sequence in which conditions occur, or the common structure amongst conditions, or are unable to differentiate between remote and proximate conditions, but would like to analyze how the impact of conditions have changed over time and cases. This approach was developed and demonstrated by Ragin and Fiss (2019) in his QCA workshop and can be defined as "set-theoretic approaches to change" Ragin and Fiss, Southern California QCA (Workshop 2019). I have termed this approach as *Relevant Variation Panel*.

According to Ragin and Fiss (2019), this approach is important because it can help us maintain the assumptions of QCA, based on studying qualitative change, while also analyzing this change over time.

To do this, researchers need to calculate the difference in fuzzy-set membership, over two or more time-points. For example, in my research, these

DOI: 10.1201/9781003384595-8

two time-points can be 1999 (as the first year included), and 2008 (as the last year studied). I have included 2008 as my second time-point, rather than 2007, because it helps me maintain contextual variation, for the set/condition "Presence of Microfinance Institutions" (MFISR). Croatia had closed all its microfinance institutions by 2008, while Montenegro had just started applying microfinance as a policy of post-conflict economic development for women. However, researchers need to *theoretically* justify the years that they have selected to study.

To calculate the difference in fuzzy-set membership, we need to follow four main steps:

1. Calibrate fuzzy-set/crisp-set membership for each variable in the first time-point.
2. Calibrate fuzzy-set/crisp-set membership for each variable in the second time-point.
3. Calculate the difference in the two fuzzy-set/crisp-set memberships for these two time-points by subtracting the first time-point from the second time-point.
4. Calibrate this difference by *defining a positive/negative change* that the researcher wants to analyze.

For example, for the variable "economic growth," measured through GDP, I defined each of the fuzzy-set calibrations shown below:

Step 1: "Set of Countries with high GDP for the year 1999"
Step 2: "Set of Countries with high GDP for the year 2008"
Step 3: Calculate the difference between "Set of Countries with high GDP for the year 2008", and "Set of Countries with high GDP for the year 1999"
Step 4: "Set of Countries with high positive change in GDP" *(meaning positive change in GDP will be given a higher calibration score, as compared to low/negative change in GDP)*

ASSUMPTIONS

The assumptions of this approach are the same as that of cross-sectional QCA, but the process of calibration is a little bit different. To analyze qualitative change over time, researchers need to define each of the sets that they are creating, including whether they are calibrating high or low change in the final/third step. In this approach, researchers need to structure their data in a wide format.

CALIBRATION PROCESS

Step 1: Calibrate each variable for time-point 1 (year 1999)

The first step in this approach is to calibrate each variable for time-point 1 (Ragin and Fiss 2019). To demonstrate this, I have applied the year 1999 as time-point 1, and calibrated each of the variables, in Table 8.1.

Step 2: Calibrate each variable for time-point 2 (year 2008)

The second step in this approach is to calibrate each variable for time-point 2. To demonstrate this, I applied the year 2008 as the second time-point, and calibrated each of the variables, as seen in Table 8.2.

Step 3: Calculate the difference in fuzzy-set membership between the two time-points

In step 3, researchers need to calculate the difference in fuzzy-set membership between the two time-points. To do this, I subtracted the difference in fuzzy-set membership for the year 1999 from the year 2008.

Step 4: Calibrate the difference in fuzzy-set membership between the two time-points

The fourth step is to calibrate the difference in fuzzy-set membership between the two time-points. This will help us analyze the qualitative change in variables, over time.

To do this, researchers need to *define and specify the direction of change* that they are calibrating. For example, I have calibrated high positive change in my conditions and outcome (Table 8.3). This is because theoretically, a high positive change in my conditions will lead to an increase in economic participation of women, at the level of the national economy, which is my outcome variable. I have added the codes for each of these steps in Table 8.4.

TABLE 8.1 Calibration Strategy for the Year 1999

CONCEPT	DIMENSION/ CHARACTERISTIC	INDICATOR/ VARIABLE	ABBREVIATION	TYPE OF SET	TYPE OF CALIBRATION	SET LABEL
Microfinance Institutions	Percentage of microfinance institutions	Number of microfinance institutions **Source:** World Bank Databank| MIX Market	FEMBR	Crisp()	threshold=1	Set of countries with MFIs present in 1999
Access to Political Freedom	The extent of democracy within the country	Democracy Scale (DEM), **Source:** Polity IV Dataset	DEMB	Crisp ()	Crisp-set (threshold=0)	Set of countries with high extent of democracy in 1999
	Access to political and civil rights	Political Rights & Civil Liberty Scale (PCL) **Source:** Freedom Rating Scale, Freedom House Reports	PCLB	Crisp ()	Crisp-set (5=0, else=1)	Set of countries with high access to political and civil rights in 1999
Access to Economic Facilities	Economic growth –> increases opportunities available to individuals for production, consumption, and exchange	Gross Domestic Product **Source:** World Bank	GDPB	Fuzzy()	Direct Calibration Method (thresholds, "e=4600000000, c=15000000000, i=23000000000")	Set of countries with high extent of economic growth in 1999

(Continued)

CONCEPT	DIMENSION/ CHARACTERISTIC	INDICATOR/ VARIABLE	ABBREVIATION	TYPE OF SET	TYPE OF CALIBRATION	SET LABEL
Access to Social Opportunities	Access to basic education and healthcare, which will help an individual live a better life	% of literate females ages 15–24 **Source:** World Bank ------- Life expectancy at birth for females, measured in years **Source:** World Bank	LTCYFB ------- LIFEFB	Fuzzy()	Direct Calibration Method (thresholds= e=98.70, c=99.00, i=99.60) ------- Direct Calibration Method (thresholds: e=76.10, c=76.30, i=76.58)	Set of countries with a high percentage of literate females in 1999 ------- Set of countries with high female life expectancy in 1999
Access to Transparency Guarantees	Society operates on the basic presumption of trust. Need for openness, guarantees of disclosure, and low corruption	Control of Corruption Scale **Source:** World Governance Indicators| World Bank	FCORB	Fuzzy()	Direct Calibration Method (thresholds, e=34.00, c=45.00, i=71.00)	Set of countries with high control of corruption in 1999
Access to Protective Security	Effectiveness of public institutions in ensuring access to basic facilities like sanitation, health, and unemployment benefits	Government Effectiveness Scale **Source**: World Governance Indicators| World Bank	NFEFFECTB	Fuzzy()	Direct Calibration Method (thresholds, e=0.00, c=15.00, i=59.00)	Set of countries with high government effectiveness in 1999

International Human Rights Treaties (Case Study Condition)	Signed human rights treaties to eliminate all forms of racial and gender inequality	Signed 18 international human rights treaties **Source**: UN Human Rights Office of the High Commissioner	HRTB	Crisp()	Crisp() (threshold=8)	Set of countries with high number of international human rights treaties signed in 1999
Female Representation in Parliament (Case Study Condition)	Percentage of seats held by women in national parliaments	Proportion of seats held by women in national parliaments (%) **Source:** Millennium Development Goals, World Bank	FRPRB	Fuzzy()	Ragin's Direct Calibration ($e=11.1$, $c=15.0$, $i=20.9$)	Set of countries with high female representation in parliament in 1999
Economic Empowerment of Women (Outcome)	Economic participation of women at the level of the national economy, for the economic sectors of own-account workers (informal economy), family workers (informal economy), employees (formal economy), employers (formal economy)	Number of women employed belonging to the age group, 15–65 years. **Source:** Employment by Sex and Status in Employment (KILM 3); ILO Modeled Estimates – Annual, calculated in thousands	OWNB, FAMB, EMPEB, EMPRB	Fuzzy()	Direct Calibration Method OWNB: (thresholds=" $e=5000$, $c=50000$, $i=80000$") FAMB: (thresholds= " $e=2000$, $c=61000$, $i=68000$") EMPEB: (thresholds= " $e=55000$, $c=260000$, $i=570000$") EMPRB: (thresholds= " $e=75000$, $c=300000$, $i=640000$")	Set of countries with high number of female own-account workers in 1999 Set of countries with high number of female family workers in 1999 Set of countries with high number of female employees in 1999 Set of countries with high number of female Employers in 1999

TABLE 8.2 Calibration Strategy for the Year 2008

CONCEPT	DIMENSION/ CHARACTERISTIC	INDICATOR/VARIABLE	ABBREVIATION	TYPE OF SET	TYPE OF CALIBRATION	SET LABEL
Microfinance Institutions	Presence of microfinance institutions	Presence of microfinance institutions **Source:** World Bank Databank|MIX Market	MFISRB	Crisp()	Threshold=1	Set of countries with presence of MFIs in 2008
Access to Political Freedom	The extent of democracy within the country	Democracy Scale (DEM) **Source:** Polity IV Dataset	DEMC	Fuzzy()	Direct Calibration Method	Set of countries with high extent of democracy in 2008
	Access to political and civil rights	Political Rights & Civil Liberty Scale (PCL) **Source:** Freedom Rating Scale, Freedom House Reports	PCLC	Fuzzy()	(thresholds="e=-66, c=1, i=9") Direct Calibration Method (thresholds="i=2.0, c=2.6, e=3.5")	Set of countries with high access to political and civil rights in 2008
Access to Economic Facilities	Economic growth –> increases opportunities available to individuals for production, consumption, and exchange	Gross Domestic Product **Source:** World Bank	GDPC	Fuzzy()	Direct Calibration Method (thresholds=" e=450000000, c=50000000000, i=70000000000)	Set of countries with economic growth in 2008

Access to Social Opportunities	Access to basic education and healthcare, which will help an individual live a better life	% of literate females between the ages of 15 to 24 years **Source:** World Bank	LTFB	Fuzzy() Direct Calibration Method (thresholds="e=99.30, c=99.50, i=99.80")	Set of countries with a high percentage of literate females in 2008
		Life expectancy at birth for females, measured in years **Source:** World Bank	LIFEFC	Direct Calibration Method (thresholds="e=76.70, c=77.50, i=79.60")	Set of countries with high female life expectancy in 2008
Access to Transparency Guarantees	Society operates on the basic presumption of trust. Need for openness, guarantees of disclosure, and low corruption	Control of Corruption Scale **Source:** World Governance Indicators| World Bank	NCORB	Fuzzy() Direct Calibration Method (thresholds="e=42.0, c=45.0, i=58.0")	Set of countries with high control of corruption in 2008
Access to Protective Security	Effectiveness of public institutions in ensuring access to basic facilities like sanitation, health, and unemployment benefits	Government Effectiveness Scale **Source:** World Governance Indicators| World Bank	NEFFECTC	Fuzzy() Direct Calibration Method (thresholds="e=0.0, c=15.0, i=59.0")	Set of countries with high government effectiveness in 2008
International Human Rights Treaties (Case Study Condition)	Signed human rights treaties to eliminate all forms of racial and gender inequality	Signed 18 international human rights treaties **Source:** UN Human Rights Office of the High Commissioner	HRTRC	Fuzzy() Fuzzy() (thresholds="e=-0.30, c=0.00, i=0.99")	Set of countries with high number of international human rights treaties signed in 2015

(Continued)

TABLE 8.2   (Continued)

CONCEPT	DIMENSION/ CHARACTERISTIC	INDICATOR/VARIABLE	ABBREVIATION	TYPE OF SET	TYPE OF CALIBRATION	SET LABEL
Female Representation in Parliament (Case Study Condition)	Percentage of seats held by women in national parliaments	Proportion of seats held by women in national parliaments (%) **Source:** Millennium Development Goals, World Bank	FRPRB	Fuzzy()	Direct Calibration Method (thresholds= "e=11.1, c=15.0, i=20.9")	Set of countries with high female representation in parliament in 2015
Economic Empowerment of Women (Outcome)	Economic participation of women at the level of the national economy, for the economic sectors of own-account workers (informal economy), family workers (informal economy), employees (formal economy), employers (formal economy)	Number of employed women ages 15–65 **Source:** Employment by Sex and Status in Employment (KILM 3); ILO Modeled Estimates – Annual, calculated in thousands	OWNC, FAMB, EMPEB, EMPRB	Fuzzy()	Direct Calibration Method OWNC: (thresholds, " e=5000, c=54000, i=99000") FAMC: (thresholds, "e=2000, c=30000, i=40000") EMPEC: (thresholds, "e=75000, c=300000, i=640000") EMPRC: (thresholds, "e=4500, c=15000, i=23000")	Set of countries with high number of female own-account workers in 2008 Set of countries with high number of female family workers in 2008 Set of countries with high number of female employees in 2008 Set of countries with high number of female employers in 2008

TABLE 8.3 Calibration Strategy for Difference in the Set-Membership Values

CONCEPT	DIMENSION/ CHARACTERISTIC	INDICATOR/VARIABLE	ABBREVIATION	TYPE OF SET	TYPE OF CALIBRATION	SET LABEL
Microfinance Institutions	Presence of microfinance institutions	Number of microfinance institutions **Source:** World Bank|Databank|MIX Market	MFIRSD	Fuzzy()	Direct Calibration Method (thresholds="e=-1, c=0.5, i=1")	Set of countries with high Positive Change in the Number of MFIs
Access to Political Freedom	The extent of democracy within the country	Democracy Scale (DEM), **Source:** Polity IV Dataset	DEME	Fuzzy()	Direct Calibration Method (thresholds: "e=-0.57, c=0.0, i=0.04")	Set of countries with high positive change in the extent of democracy
	Access to political and civil rights	Political Rights & Civil Liberty Scale (PCL) **Source:** Freedom Rating Scale, Freedom House Reports	PCLD	Fuzzy()	Direct Calibration Method (thresholds: "e=-0.06, c=0.08, i=0.20")	Set of countries with high positive change in the political and civil rights scale
Access to Economic Facilities	Economic growth –> increases opportunities available to individuals for production, consumption, and exchange	Gross Domestic Product **Source:** World Bank	GDPE	Fuzzy()	Direct Calibration Method (thresholds= "e=-0.720, c=0.010, i=0.060")	Set of countries with high positive change in economic growth

(Continued)

TABLE 8.3 (Continued)

CONCEPT	DIMENSION/ CHARACTERISTIC	INDICATOR/VARIABLE	ABBREVIATION	TYPE OF SET	TYPE OF CALIBRATION	SET LABEL
Access to Social Opportunities	Access to basic education and healthcare, which will help an individual live a better life	% of literate females ages 15–24 **Source:** World Bank	LTRCYFE	Fuzzy()	Direct Calibration Method (thresholds="e=−0.35, c=0.0, i=0.89")	Set of countries with high positive change in percentage of literate females
		Life expectancy at birth for females, measured in years **Source:** World Bank	LIFEXFE		Direct Calibration Method (thresholds=" e=−0.35, c=0.1, i=0.89")	Set of countries with high positive change in female life expectancy
Access to Transparency Guarantees	Society operates on the basic presumption of trust. Need for openness, guarantees of disclosure, and low corruption	Control of Corruption Scale **Source:** World Governance Indicators| World Bank	NCORD	Fuzzy()	Direct Calibration Method (thresholds, "e=−0.25, c=0.0, i=0.89")	Set of countries with high positive change in control of corruption
Access to Protective Security	Effectiveness of public institutions in ensuring access to basic facilities like sanitation, health, and unemployment benefits	Government Effectiveness Scale **Source:** World Governance Indicators| World Bank	NEFFECTE	Fuzzy()	Direct Calibration Method (thresholds, "e=−0.27, c=0.10, i=0.41")	Set of countries with high positive change in government effectiveness

Variable	Description	Measure / Source	Code	Type	Calibration	Relevant Variation
International Human Rights Treaties (Case Study Condition)	Signed human rights treaties to eliminate all forms of racial and gender inequality	Signed 18 international human rights treaties **Source:** UN Human Rights Office of the High Commissioner	HRTRD	Fuzzy()	Direct Calibration Method (thresholds = "e=-0.30, c=0.00, i=0.99")	Set of countries with high positive change in number of international human rights treaties signed
Female Representation in Parliament (Case Study Condition)	Percentage of seats held by women in national parliaments	Proportion of seats held by women in national parliaments (%) **Source:** Millennium Development Goals, World Bank	FRPRD	Fuzzy()	Direct Calibration Method (thresholds, "e=-0.05, c=0.00, i=0.08")	Set of countries with high positive change in female parliamentary representation
Economic Empowerment of Women (Outcome)	Economic participation of women at the level of the national economy, for the economic sectors of own-account workers (informal economy), family workers (informal economy), employees (formal economy), employers (formal economy)	Number of employed women ages 15–65 **Source:** Employment by Sex and Status in Employment (KILM 3); ILO Modeled Estimates – Annual, calculated in thousands	OWND, FAMD, EMPEF, EMPRD	Fuzzy()	Direct Calibration Method OWND: (thresholds, "e=-0.027, c=0.001, i=0.002") FAMD: (thresholds, "e=-0.500, c=0.200, i=0.400") EMPEF: (thresholds, "e=-0.0015, c=-0.0007, i=-0.0004") EMPRD: (thresholds, "e=-0.0007, c=0.00955, i=0.0096")	Set of countries with high positive change in number of female own-account workers / Set of countries with high positive change in number of female family workers / Set of countries with high positive change in number of female employees / Set of countries with high positive change in number of female employers

TABLE 8.4   Codes for Steps 1–4, Set of Countries with High Positive Change in the number of MFIs

```
str(DATA$PresenceofMFIs_1999)
Xplot(DATA$PresenceofMFIs_1999, jitter=TRUE)
DATA$MFISR<-calibrate(DATA$PresenceofMFIs_1999,
 type="crisp", thresholds=1)
skew.check(DATA$MFISR)
str(DATA$PresenceofMFIs_2008)
Xplot(DATA$PresenceofMFIs_2008, jitter=TRUE)
DATA$MFISRB<-calibrate(DATA$PresenceofMFIs_2008,
 type="crisp", thresholds=1)
skew.check(DATA$MFISRB)
DATA$MFISRC<-DATA$MFISRB-DATA$MFISR
str(DATA$MFISRC)
DATA$MFIRSD<-calibrate(DATA$MFISRC,type="fuzzy",threshold
 s="e=-1,c=0.5, i=1", logistic = TRUE)
skew.check(DATA$MFISRD)
```

As mentioned before, in Table 8.4, I have applied the necessary condition, "Presence of Microfinance Institutions", as an example to demonstrate the codes for all the four steps described above. I have defined the set as a "Set of countries with high Positive Change in the Number of MFIs." I have included the codes for my other conditions, as my appendix material, at the end of the chapter.

Once we have calibrated the qualitative change in all the variables, we can then start testing for necessary and sufficient conditions. In the next section, I first demonstrate how to test for necessary conditions, and I then demonstrate how to test for sufficient conditions.

TESTING FOR NECESSARY CONDITIONS

For this model, I have applied "Economic Participation of Women as Own-Account Workers," OWNE, as the outcome variable.

My testing for necessary conditions showed that High Government Effectiveness, NEFFECTE, is a necessary condition for Economic Participation of Women as Own-Account Workers, OWNE, Table 8.6. But the absence of High Government Effectiveness (~NEFFECTE) did not

TABLE 8.5 Code for Testing of "High NEFFECTE" as a Necessary Condition

```
QCAfit(PANELF$NEFFECTE, PANELF$OWNE, cond.lab="High
 NEFFECT", necessity = TRUE, consH = TRUE)
QCAfit(1-PANELF$NEFFECTE, PANELF$OWNE, cond.lab="Absence
 of High NEFFECT", necessity = TRUE, consH = TRUE)
```

TABLE 8.6 Data Analysis Results for Testing of "High NEFFECTE" as a Necessary Condition

CONDITION	CONS.NEC	COV.NEC	RON
High NEFFECT	0.978	0.999 1	
Absence of High NEFFECT	0.321	0.255	0.51

lead to the absence of Economic Participation of Women as Own-Account Workers (~OWNE). Therefore, NEFFECTE is not a necessary condition for OWNE.

I have tabulated the codes from the testing of NEFFECTE as a necessary condition for OWNE, in Table 8.5, and added the other codes in my appendix material, at the end of this chapter.

TESTING FOR SUFFICIENT CONDITIONS

To test for sufficient conditions, we need to first create a truth table using the truthTable() function. As discussed in the previous chapters, the truth tables must be created for the presence and absence of the outcome separately. The second step is to minimize these truth tables to create conservative, parsimonious, and intermediate solutions. The third step is to check for errors and untenable assumptions and create enhanced conservative, parsimonious, and intermediate solutions. We also need to specify the directional expectations for the intermediate and enhanced intermediate solutions, Table 8.7.

I have tabulated the codes for each of these steps below, in Tables 8.8 to 8.13.

TABLE 8.7   Directional Expectations Between Conditions and Outcome

CONDITIONS	INCREASES ECO. PARTICIPATION	DECREASES ECO. PARTICIPATION
MFISR	Present(1)	Absent(0)
HRTR	Present(1)	Absent(0)
GDPR	Present(1)	Absent(0)
DEMC	Present(1)	Absent(0)
DPCL	Present(1)	Absent(0)
LTRCYR	Present(1)	Absent(0)
LIFEXFR	Present(1)	Absent(0)
FCORR	Present(1)	Absent(0)
NEFFECTR	Present(1)	Absent(0)
FRPR	Present(1)	Absent(0)

TABLE 8.8   Code for Creating the Truth Tables

```
# Presence of outcome:
ttF<-truthTable(data=PANELF, outcome="OWNE",
conditions= c("MFIRSD", "GDPE", "DEMD", "PCLE",
 "LTRCYFE", "LIFEXFE", "NCORD", "NEFFECTE", "FRPRD",
 "HRTRD"), incl.cut = 0.85, sort.by = "incl,n", pri.
 cut=0.51, dcc= TRUE, decreasing=FALSE, complete=TRUE,
 show.cases = TRUE)
ttF
# Absence of outcome:
ttf<-truthTable(data=PANELF, outcome="~OWNE",
conditions= c("MFIRSD", "GDPE", "DEMD", "PCLE",
 "LTRCYFE", "LIFEXFE", "NCORD", "NEFFECTE", "FRPRD",
 "HRTRD"), incl.cut = 0.85, sort.by = "incl,n", pri.
 cut=0.51, dcc= TRUE, decreasing=FALSE,complete=TRUE,
 show.cases = TRUE)
ttf
```

TABLE 8.9   Code for Boolean Minimization

```
sol_F<-minimize(input=ttF, details=TRUE, row.dom=TRUE)
sol_F
sol_f<-minimize(input=ttf, details=TRUE, row.dom=TRUE)
sol_f
```

TABLE 8.10 Code for the Conservative, Parsimonious, and Intermediate Solutions, Presence of the Outcome

```
sol_CF <- minimize(ttF, details=TRUE, row.dom=TRUE)
sol_CF
sol_PF <-minimize(ttF, include="?", details=TRUE, row.
 dom=TRUE)
sol_PF
sol_IF <- minimize(ttF, include = "?", dir.exp = "1, 1,
 1, 1, 1, 1, 1, 1, 1, 1",details=TRUE, row.dom=TRUE)
sol_IF
```

TABLE 8.11 Code for the Conservative, Parsimonious, and Intermediate Solutions, Absence of the Outcome

```
sol_cf <- minimize(ttf, details=TRUE, row.dom=TRUE)
sol_cf
sol_pf<-minimize(ttf, include="?", details=TRUE, row.
 dom=TRUE)
sol_pf
sol_if <- minimize(ttf, include = "?", dir.exp = "0, 0,
 0, 0, 0, 0, 0, 0, 0, 0",details=TRUE, row.dom=TRUE)
sol_if
```

TABLE 8.12 Code for Identifying Errors

```
TYPETWO<-findRows(obj=ttF, type=2)
TYPETWO
findRows(obj=ttF, type=3)
```

TABLE 8.13 Codes for Creating Enhanced Conservative, Parsimonious, and Intermediate Solutions

```
sol_CFN<-minimize(ttF, details=TRUE, exclude=c(TYPETWO),
 row.dom=TRUE)
sol_CFN
sol_PFN<-minimize(ttF, include="?", details=TRUE,
 exclude=c(TYPETWO), row.dom=TRUE)
sol_PFN
sol_IFN<-minimize(ttF, include = "?", dir.exp = "1, 1, 1,
 1, 1, 1, 1, 1, 1, 1",details=TRUE, exclude=c(TYPETWO),
 row.dom=TRUE)
sol_IFN
```

TABLE 8.14 The Enhanced Intermediate Solution

```
From C1P1:

M1:    MFIRSD*PCLE*NEFFECTE*FRPRD*HRTRD  -> OWNE

                                 inclS   PRI    covS   covU  cases
-----------------------------------------------------------------------
1  MFIRSD*PCLE*NEFFECTE*FRPRD*HRTRD  0.999  0.999  0.500    -      3
-----------------------------------------------------------------------
                            M1  0.999  0.999  0.500
```

INTERPRETING THE ENHANCED INTERMEDIATE SOLUTION

The interpretation of the enhanced intermediate solution is similar to Chapters 6 and 7.

As an example, my enhanced intermediate solution shows that a high positive change in the number of microfinance institutions (MFISRD), combined with high positive change in factors like political and civil liberties (PCLE), government effectiveness (NEFFECTE), female representation in parliament (FRPRD), and international human rights treaties signed (HRTRD), increased the economic participation in the case of Montenegro, over the years 1999 to 2008. The overall solution consistency is 0.99, and the overall coverage is 0.50 (Table 8.14).

VISUALIZING THE PATHWAYS

To visualize the pathways from the testing of necessary and sufficient conditions, we can apply the XYplot() or the xy.plot() functions. I have tabulated the codes, in Table 8.15, and visualized the pathway for the enhanced intermediate solution, in Figures 8.1 and 8.2.

TABLE 8.15   Code for Visualizing the Solution Pathways

```
PANELF$Observations<-NULL
XYplot(NEFFECTE, OWNE, data=PANELF, relation =
 "necessity", enhance = TRUE, jitter=TRUE, clabels =
 seq(nrow(PANELF)))
XYplot(MFIRSD*PCLE*NEFFECTE*FRPRD*HRTRD, OWNE,
 data=PANELF, relation = "sufficiency", enhance = TRUE,
 jitter=TRUE, clabels = seq(nrow(PANELF)))
```

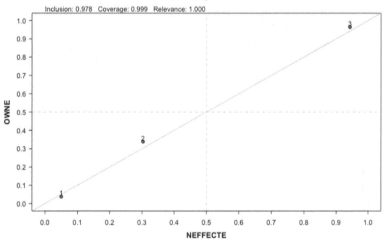

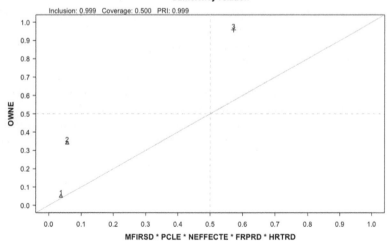

STRENGTHS AND WEAKNESSES

In this chapter, I discussed Relevant Variation Panel as a fourth approach toward analyzing panel data in QCA.

This approach is important because:

- It can help us analyze relevant variation (qualitative change) over time, rather than increasing the number of cases studied, by measuring each case, country_year, repeatedly in a panel data format, as discussed in Chapter 5.
- This approach can also help us how a qualitative change in conditions (independent variables) can impact the outcome variable (dependent variable/s), as compared to the Sub-QCA Model approach discussed in Chapter 6.
- Researchers can also set up multiple time-points, rather than just analyzing the first and last years of observation, but they should theoretically justify their selection of years.

Despite these strengths, there are two important drawbacks of *Relevant Variation Panel*:

First, we need to maintain a balance between the number of variables and the number of observations/cases studied.

Second, researchers might have to choose a different time-point or omit those variables that do not show contextual variation at a certain time-point. In this case, a researcher can include a different time-point/year or not include the variable at all. For example, my case studies did not start applying for European Union membership in 1999, so I could not calibrate this variable for that year, as all the observations were zero. To enable comparison among the different QCA approaches discussed in this book, I excluded this variable. I applied a similar strategy to include 2008 as my second time-point (rather than the year 2007), as my data did not show contextual variation for my condition "Presence of Microfinance Institutions."

In the next chapter, I will conclude this book by briefly summarizing and comparing the strengths and weaknesses of the four panel data approaches that I have discussed.

Appendix

- Dataset: FinalData_1999_2008.csv

OBSERVATIONS	PRESENCEOFMFIS_1999	PRESENCEOFMFIS_2008
Bosnia and Herzegovina	1	1
Croatia	1	0
Montenegro	0	1

OBSERVATIONS	NUMBEROFMFIS_1999	NUMBEROFMFIS_2008
Bosnia and Herzegovina	18	18
Croatia	3	0
Montenegro	4	4

OBSERVATIONS	GDP_1999	GDP_2008	DEM_1999	DEM_2008	PCL_1999
Bosnia and Herzegovina	4.69E+09	1.91E+10	−66	−66	5
Croatia	2.34E+10	7.03E+10	0	0	4
Montenegro	1.84E+10	4.55E+09	0	9	5

OBSERVATIONS	PCL_2008	LTRCYF_1999	LTRCYF_2008	LIFEXF_1999
Bosnia and Herzegovina	3.5	98.73545	99.84363	76.581
Croatia	2	99.6044	99.65321	76.55
Montenegro	3	98.98921	99.3606	76.11

OBSERVATIONS	LIFEXF_2008	FCOR_1999	FCOR_2008	FEFFECT_1999
Bosnia and Herzegovina	78.182	NA	42.71844	NA
Croatia	79.6	NA	58.25243	NA
Montenegro	76.755	NA	51.94175	NA

OBSERVATIONS	FEFFECT_2008	FPARL_1999	FPARL_2008	EUM_1999
Bosnia and Herzegovina	33.49514	NA	11.9	0
Croatia	71.35922	7.9	20.9	0
Montenegro	53.8835	NA	11.1	0

OBSERVATIONS	EUM_2008	INTHRT_1999	INTHRT_2008	OWN_1999
Bosnia and Herzegovina	1	8	13	49876
Croatia	2	8	14	82003
Montenegro	1	0	11	5022

OBSERVATIONS	OWN_2008	FAM_1999	FAM_2008	EMPE_1999
Bosnia and Herzegovina	52019	59102	40520	241554
Croatia	99998	68965	28527	574492
Montenegro	5734	2040	2394	55634

OBSERVATIONS	EMPE_2008	EMPR_1999	EMPR_2008	COUNTRY	YEAR
Bosnia and Herzegovina	279573	10745	13646	BiH	1999
Croatia	641158	24269	23820	HRV	1999
Montenegro	75007	3255	4510	MNE	1999

Codes for testing of necessary conditions, presence

```
QCAfit(PANELF$MFIRSD, PANELF$OWNE, cond.lab="Presence
of MFIs", necessity = TRUE, consH = TRUE)
QCAfit(PANELF$GDPE, PANELF$OWNE, cond.lab="High GDP",
necessity = TRUE, consH = TRUE)
QCAfit(PANELF$DEMD, PANELF$OWNE, cond.lab="High DEMC",
necessity = TRUE, consH = TRUE)
QCAfit(PANELF$PCLE, PANELF$OWNE, cond.lab="High DPCL",
necessity = TRUE, consH = TRUE)
QCAfit(PANELF$LTRCYFE, PANELF$OWNE, cond.lab="High
LTRCYR", necessity = TRUE, consH = TRUE)
QCAfit(PANELF$LIFEXFE, PANELF$OWNE, cond.lab="High
LIFEXFR", necessity = TRUE, consH = TRUE)
```

```
QCAfit(PANELF$NCORD, PANELF$OWNE, cond.lab="High
NCORR", necessity = TRUE, consH = TRUE)

QCAfit(PANELF$NEFFECTE, PANELF$OWNE, cond.lab="High
NEFFECT", necessity = TRUE, consH = TRUE)

QCAfit(PANELF$FRPRD, PANELF$OWNE, cond.lab="High FRPR",
necessity = TRUE, consH = TRUE)

QCAfit(PANELF$HRTRD, PANELF$OWNE, cond.lab="High HRTR",
necessity = TRUE, consH = TRUE)
```

Codes for testing of necessary conditions, absence

```
QCAfit(1-PANELF$MFIRSD, PANELF$OWNE, cond.lab="Absence
of MFIs", necessity = TRUE, consH = TRUE)

QCAfit(1-PANELF$GDPE, PANELF$OWNE, cond.lab="Absence
of High GDP", necessity = TRUE, consH = TRUE)

QCAfit(1-PANELF$DEMD, PANELF$OWNE, cond.lab="Absence
of High DEMC", necessity = TRUE, consH = TRUE)

QCAfit(1-PANELF$PCLE, PANELF$OWNE, cond.lab="Absence
of High DPCL", necessity = TRUE, consH = TRUE)

QCAfit(1-PANELF$LTRCYFE, PANELF$OWNE, cond.lab="Absence
of High LTRCYR", necessity = TRUE, consH = TRUE)

QCAfit(1-PANELF$LIFEXFE, PANELF$OWNE, cond.lab="Absence
of High LIFEXFR", necessity = TRUE, consH = TRUE)

QCAfit(1-PANELF$NCORD, PANELF$OWNE, cond.lab="Absence
of High NCORR", necessity = TRUE, consH = TRUE)

QCAfit(1-PANELF$NEFFECTE, PANELF$OWNE, cond.
lab="Absence of High NEFFECT", necessity = TRUE, consH
= TRUE)

QCAfit(1-PANELF$FRPRD, PANELF$OWNE, cond.lab="Absence
of High FRPR", necessity = TRUE, consH = TRUE)

QCAfit(1-PANELF$HRTRD, PANELF$OWNE, cond.lab="Absence
of High HRTR", necessity = TRUE, consH = TRUE)
```

Conclusion

9

In this book, I have demonstrated four different approaches that researchers can apply while analyzing panel data in QCA. Researchers can adopt either of these approaches since they all produce quite similar data analysis results, but I have highlighted a few suggestions below. Researchers can also adopt one or more of these approaches to cross-check their data analysis results, which will help them address robustness in panel data QCA.

I encourage researchers to:

- Adopt a "Multiple Sub-QCA" approach, as discussed in Chapter 6, if they already know the years that might turn out to be significant. Researchers can also apply this approach as a second step in multi-method research, as it will help researchers maintain a balance between the number of cases and the number of conditions.
- Adopt a "Remote-Sufficiency Panel" approach, as discussed in Chapter 7, if they can theoretically differentiate between remote and proximate conditions, or if their case studies reveal remote and proximate conditions. Along with this, researchers need to apply the cluster() function to see if the impact of conditions varies over time and across the cases studied. Remember to meet the requirements for the consistency, coverage, and RoN thresholds, for remote SUIN conditions.
- Adopt a "Relevant Variation Panel" approach, as discussed in Chapter 8, if they can maintain a balance between the number of conditions and the number of cases studied. Hence, it might be better for researchers to adopt this approach as a second step in multi-method research.
- Adopt a "Cluster QCA" approach, discussed in Chapter 5, as a first step in multi-method research analysis. Since this approach might overemphasize the impact of conditions that are repeatedly measured across multiple observations (country_years), researchers should follow it up with a set-theoretic multi-method research (SMMR), which will help them better understand why certain cases

DOI: 10.1201/9781003384595-9

turned out to be typical, and to interpret the pathways in the context of the cases studied.

- Structure their data in a long or a wide format, depending on the approach that they are adopting. For *"Cluster QCA"* (Chapter 5), *"Multiple Sub-QCA, Model 2"* (Chapter 6), and *"Remote-Proximate Panel"* (Chapter 7), researchers need to structure their data in a long format, while for *"Multiple Sub-QCA, Model 1"* (Chapter 6) and *"Relevant Variation QCA"* (Chapter 8), researchers need to organize their data in a wide format.
- Calibrate their data separately for each approach that they are planning to adopt.

References

WORKS CITED

Basurto, X., & Speer, J. (2012). "Structuring the Calibration of Qualitative Data as Sets for Qualitative Comparative Analysis (QCA)." *Field Methods* 24, no. 2: 155–174. https://doi.org/10.1177/1525822X11433998

Baumgartner, Michael. 2009. "Inferring Causal Complexity." *Sociological Methods & Research* 38, no. 1: 71–101.

Bhattacharya, Preya. 2020. "Can Microfinance Impact National Economic Development? A Gendered Perspective." Doctoral Dissertation. Available at: http://rave.ohiolink.edu/etdc/view?acc_num=kent1597080132873571 (Accepted for publication with Bloomsbury Publishing, in 2023)

Bhattacharya, Preya. 2023a. How to Build and Analyze a Panel Data QCA Model? A Methodological Demonstration of Garcia-Castro and Ariño's Panel Data QCA Model. *Methodological Innovations* 16, no. 3: 265–275. https://doi.org/10.1177/20597991231179389

Bhattacharya, Preya. 2023b. "What Is Ragin's Indirect Method of Calibration?" *International Journal of Social Research Methodology* 26, no. 6: 825–831. https://doi.org/10.1080/13645579.2022.2110732

Boole, George. 1854, 2000. *An Investigation of the Laws of Thought: On Which Are Founded the Mathematical Theories of Logic and Probabilities of Cambridge Library Collection – Mathematics.* Cambridge: Cambridge University Press. Available at: https://doi.org/10.1017/CBO9780511693090

Boole, George. 1847, 2009. *The Mathematical Analysis of Logic: Being an Essay Towards a Calculus of Deductive Reasoning.* Cambridge Library Collection – Mathematics. Cambridge: Cambridge University Press. doi: 10.1017/CBO9780511701337.

Caren, Neal, and Aaron Panofsky. 2005. "TQCA: A Technique for Adding Temporality to Qualitative Comparative Analysis." *Sociological Methods & Research* 34: 147–172.

Dusa, Adrian. 2019. *QCA with R: A Comprehensive Resource.* (R package). Springer International Publishing.

Dusa, Adrian. 2022. *QCA with R: A Comprehensive Resource.* Available at: https://bookdown.org/dusadrian/QCAbook/QCAbook.pdf

Emma Uprichard, Rosalind Edwards & Malcolm Williams. (2018). IJSRM's 100th issue. International Journal of Social Research Methodology 21, no. 1: 1–6.

European Union. *Country Profiles*. Available at: https://europa.eu/european-union/ about-eu/countries_en#joining (accessed January 7, 2024).

Freedom House. 2020. Freedom in the World: A Leaderless Struggle for Democracy. Freedom in the World Dataset. Country and Territory Ratings and Statuses, 1973–2023. Available at: https://freedomhouse.org/report/freedom-world#Data

Freedom House Report. Available at: https://freedomhouse.org/reports/freedom-world/ freedom-world-research-methodology

Furnari, Santi. 2018. Four approaches to longitudinal QCA: Opportunities and challenges. Presentation at the QCA PDW, Academy of Management 2018, Chicago.

Garcia-Castro, Roberto, and Miguel A. Ariño. 2016. "A General Approach to Panel Data Set-Theoretic Research." *Journal of Advancement in Management Sciences & Information Systems*, 2: 63–76.

Goertz, Gary. 2006. *Social Science Concepts: A Users Guide*. Princeton University Press.

Goertz, Gary, and James Mahoney. 2012. *A Tale of Two Cultures: Qualitative and Quantitative Research in the Social Sciences*. Princeton, NJ: Princeton University Press.

International Labor Organization. ILOSTAT explorer. Available at: https://www.ilo.org/ shinyapps/bulkexplorer4/?lang=en&segment=indicator&id=EMP_2EMP_SEX_ STE_NB_A (accessed January 7, 2024).

Lijphart, Arendt. 1975. "The Comparable Cases Strategy in Comparative Research." *Comparative Political Studies* 8, no.2 (July): 158–177. https://doi.org/10.1177/ 001041407500800203.

Mahoney, James, and Dichter Reuschmeyer. 2003. Comparative historical analysis: Achievements and agendas. In J. Mahoney and D. Reuschmeyer (Eds.), *Comparative Historical Analysis in the Social Sciences*. Cambridge: Cambridge University Press.

Mello, Patrick A. 2021. *Qualitative Comparative Analysis: An Introduction to Research Design and Application*. Washington, DC: Georgetown University Press.

Oana, Ioana-Elena, and Carsten Q. Schneider. 2018. "Set Methods: An Add-on Package for Advanced QCA". *The R Journal* 10, no. 1: 507–533. https://journal.r-project. org/archive/2018/RJ-2018-031/RJ-2018-031.pdf

Oana, Ioana-Elena, Carsten Q. Schneider, and Eva Thomann. 2021. *Qualitative Comparative Analysis Using R: A Beginner's Guide*. Cambridge: Cambridge University Press.

Ragin, Charles C. 2000. *Fuzzy-Set Social Science*. Chicago and London: The University of Chicago Press.

Ragin, Charles C. 1987, 2014. *The Comparative Method: Moving Beyond Qualitative and Quantitative Strategies*. California: University of California Press.

Ragin, Charles C. 2008. *Redesigning Social Inquiry: Fuzzy Sets and Beyond*. Chicago: The University of Chicago Press.

Ragin, Charles C., and Peer C. Fiss. 2019. *Southern California QCA Workshop*. Irvine: University of California.

Rihoux, Benoit, and Bojana Lobe. 2009. The case for Qualitative Comparative Analysis (QCA): Adding leverage for thick cross-case comparison. In D. Byrne and C. Ragin (Eds.), *The Sage Handbook of Case-Based Methods*. London: Sage Publications Ltd. https://dx.doi.org/10.4135/9781446249413

Rihoux, Benoit, and Gisele De Meur. 2009. Crisp-Set Qualitative Comparative Analysis. In R. Benoit and C. Ragin (Eds.), *Configurational Comparative Methods: Qualitative Comparative Analysis (QCA) and Related Techniques*. Thousand Oaks, California: Sage Publications, Inc.

Rohlfing, I., and Schneider, C. Q. (2018). "A Unifying Framework for Causal Analysis in Set-Theoretic Multimethod Research." *Sociological Methods & Research*, 47, no. 1: 37–63. https://doi.org/10.1177/0049124115626170

Schneider, C. Q., & Rohlfing, I. (2013). "Combining QCA and Process Tracing in Set-Theoretic Multi-Method Research." *Sociological Methods & Research*, 42, no. 4: 559–597. https://doi.org/10.1177/0049124113481341

Schneider, Carsten Q. 2019. "Two-step QCA Revisited: The Necessity of Context Conditions." *Quality and Quantity* 53: 1109–1126. https://doi.org/10.1007/s11135-018-0805-7

Schneider, Carsten Q., and Claudius Wagemann. 2006. "Reducing Complexity in Qualitative Comparative Analysis (QCA): Remote and Proximate Factors and the Consolidation of Democracy." *European Journal of Political Research* 75: 751–786. https://ejpr.onlinelibrary.wiley.com/doi/abs/10.1111/j.1475–6765.2006.00635.x.

Schneider, Carsten Q., and Claudius Wagemann. 2012. *Set-Theoretic Methods for the Social Sciences: A Guide to Qualitative Comparative Analysis*. New York: Cambridge University Press.

Schneider, Carsten Q., and Ingo Rohlfing. 2013. "Combining QCA and Process Tracing in Set-Theoretic Multi-Method Research." *Sociological Methods & Research* 42, no. 4: 559–597. https://doi.org/10.1177/0049124113481341

Skocpol, Theda, and Margaret Sommers. 1980. "The Uses of Comparative History in Macrosocial Inquiry." *Comparative Studies in Society and History* 22, no. 2 (April): 174–197. Available at: www.jstor.org/stable/178404?origin=JSTOR-pdf

United Nations Human Rights, Office of The High Commissioner. Ratification of 18 International Human Rights Treaties. Available at: https://indicators.ohchr.org/ (accessed January 7, 2024).

Verweij, Stephan, and Barbara Vis. 2021. "Three Strategies to Track Configurations Over Time with Qualitative Comparative Analysis." *European Political Science Review* 13: 95–111. https://doi.org/10.1017/S1755773920000375

World Bank, Data Bank. MIX Market. Available at: https://databank.worldbank.org/source/mix-market (accessed January 7, 2024).

World Bank, Data. GDP (current US$). Available at: https://data.worldbank.org/indicator/NY.GDP.MKTP.CD?view=chart (accessed January 7, 2024).

World Bank, Data. Literacy Rate, Youth Female (% of female ages 15–24). Available at: https://data.worldbank.org/indicator/SE.ADT.1524.LT.FE.ZS (accessed January 7, 2024).

World Bank, Data. Life Expectancy at Birth, female (years). Available at: https://data.worldbank.org/indicator/SP.DYN.LE00.FE.IN (accessed January 7, 2024).

World Bank, Data Bank. Worldwide Governance Indicators. Available at: https://databank.worldbank.org/source/worldwide-governance-indicators (accessed January 7, 2024).

World Bank, Data Bank. Millennium Development Goals. Available at: https://databank.worldbank.org/source/millennium-development-goals (accessed January 7, 2024).

Zadeh, Lotfi A. 1965. "Fuzzy Sets". *Information and Control* 8: 338–353. https://reader.elsevier.com/reader/sd/pii/S001999586590241X?token=2D35D982A5143A47BAC3A7BFA166CA67C94C9CAD6C62B7440C5951A47A5282952D854700995F92A7620C30B7B3AF9DE5&originRegion=us-east-1&originCreation=20221127185707

Index

Note: Page numbers in **bold** indicate tables on the corresponding pages.